Student Solutions

to accompany

Finite Mathematics

fourth edition

Howard Rolf
Baylor University

Saunders College Publishing
Harcourt Brace College Publishers

Fort Worth Philadelphia San Diego New York Orlando Austin
San Antonio Toronto Montreal London Sydney Tokyo

Printed in the United States of America

ISBN 0-03-022073-4

901 129 76543

Preface

This Student's Solution Manual accompanies **Finite Mathematics, Fourth Edition** by Howard L. Rolf. All references to chapters, sections, and exercises refer to the textbook. This manual contains worked out solutions to the odd-numbered exercises in the textbook.

Do not use this manual as a substitute for working the homework yourself. You should use this manual as a last resort in solving a problem.

You may find it useful to compare your solution with the solution in this manual. Be aware that your solution may be different, yet correct. In fact, you may have a clever solution that is better. Quite often the steps to a solution may occur in a different order and still be correct. Therefore, do not assume your solution is incorrect if it differs from the one in this manual.

Contents

Chapter 1
Functions and Lines

Section 1.1

1. $y = 15x + 20$ Domain is number of hours worked. Range is number of dollars of fee.

3. (a) $f(5) = \$23.75$ (b) $f(3) = \$14.25$

5. (a) $f(1) = 4(1) - 3 = 4 - 3 = 1$
    (b) $f(-2) = 4(-2) - 3 = -8 - 3 = -11$
    (c) $f(1/2) = 4(1/2) - 3 = 2 - 3 = -1$ (d) $f(a) = 4a - 3$

7. (a) $f(5) = \dfrac{5 + 1}{5 - 1} = \dfrac{6}{4} = \dfrac{3}{2}$ (b) $f(-6) = \dfrac{-6 + 1}{-6 - 1} = \dfrac{5}{7}$

    (c) $f(0) = \dfrac{0 + 1}{0 - 1} = \dfrac{1}{-1} = -1$ (d) $f(2c) = \dfrac{(2c + 1)}{(2c - 1)}$

9. (a) $C(5) = 78(5) = 390$; $C(2.5) = 78(2.5) = 195$;
         $C(6.4) = 78(6.4) = 499.2$
    (b) Solve $78x = 741$ $x = 9.5$ ounces

11. (a) $f(60) = 9(60) = 540$ calories
    (b) Solve $9x = 750$ $x = 83.3$ minutes

13. Let x = number of hamburgers. $y = 1.80x + 25$

15. Let x = regular price. $y = x - .20x$ or $y = 0.80x$

17. Let x = number of loads. $y = 0.60x + 12$

19. Let x = number of students. $y = 3500x + 5,000,000$

21. Let x = list price. $y = 0.88x$

23. (a) $A(450) = 30(450) = 13,500 \text{ ft}^2$
    (b) $A(125) = 30(125) = 3,750 \text{ ft}^2$
    (c) $0.4A(650) = 30(650)(0.4) = 19,500(0.4) = \$7,800$
    (d) Solve $30w = 15,900$ $w = 530$ ft

25. (a) $A = \pi r^2$ is a function
    (b) Domain: positive numbers. Range: positive numbers.

27. (a) p = price per pound times w is a function
    (b) Domain: positive numbers. Range: positive numbers.

29. (a) $y = x^2$ is a function
    (b) Domain: all real numbers. Range: All nonnegative numbers.

31. y is not a function of x. There can be more than one person with a given family name. x is function of y.

33. Not a function because two classes can have the same number of boys, but the with combined weights different.

35. Not a function because two families with the same number of children can have a different number of boys.

39. Let x = age and y = pulse rate.
    (a) $p = .40(220 - x)$ (b) $p = .70(220 - x)$

41. (a) $6000 = k(30)^3 = 27000k$ gives $k = .2222$
     (b) $p = .2222(20)^3 = 1777.6$ watts
     (c)   The number of watts and wind speed are not negative.

Section 1.2

1. $f(0) = 8$, $f(1) = 11$

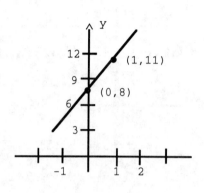

3. $f(1) = 8$, $f(-1) = 6$

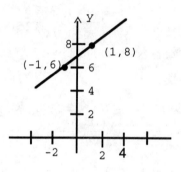

5. $f(0) = -1$, $f(-1) = 2$

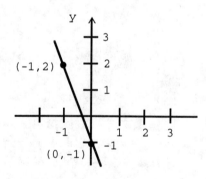

7.   Slope = 7, y-intercept = 22

9.   Slope = -2/5, y-intercept = 6

11. Solve for y
        $5y = -2x + 3$
        $y = -\dfrac{2}{5} x + \dfrac{3}{5}$
      Slope = -2/5, y-intercept = 3/5

13. Solve for y
        $3y = x + 6$
        $y = \dfrac{1}{3} x + 2$
      Slope = 1/3, y-intercept = 2

2

15. $m = \dfrac{4 - 2}{3 - 1} = \dfrac{2}{2} = 1$ 17. $m = \dfrac{-5 - (-1)}{-1 - (-4)} = -\dfrac{4}{3}$

19. $y = -2$ 21. $y = 0$

23. 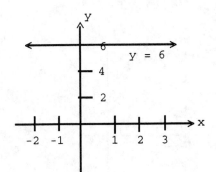   25.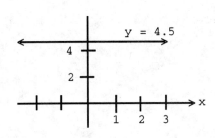

27. $m = \dfrac{5 - 2}{3 - 3} = \dfrac{3}{0}$ m is undefined so the graph is a vertical line $x = 3$

29. Vertical line $x = 10$

31. 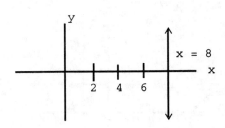   33.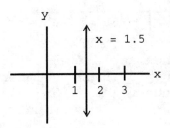

35. $y = 4x + 3$ 37. $y = -x + 6$

39. $y = \dfrac{1}{2}x$

41. $y = -4x + b$
$1 = -4(2) + b$
$b = 9$
$y = -4x + 9$

43. $y = \dfrac{1}{2}x + b$
$4 = \dfrac{1}{2}(5) + b$
$b = \dfrac{3}{2}$
$y = \dfrac{1}{2}x + \dfrac{3}{2}$

45. $y - 5 = 7(x - 1)$
$y = 7x - 7 + 5$
$y = 7x - 2$

47. $y - 6 = \dfrac{1}{5}(x - 9)$
$y = \dfrac{1}{5}x - \dfrac{9}{5} + 6$
$y = \dfrac{1}{5}x + \dfrac{21}{5}$

49. $m = \dfrac{1 - 0}{2 + 1} = \dfrac{1}{3}$ $y - 0 = \dfrac{1}{3}(x + 1)$

    $y = \dfrac{1}{3}x + \dfrac{1}{3}$

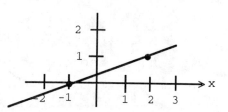

51. $m = \dfrac{2 - 0}{1 - 0} = \dfrac{2}{1} = 2$

    $y - 0 = 2(x - 0)$

    $y = 2x$

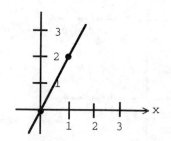

53. $y = 4$

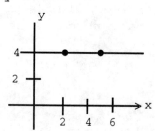

55. When $x = 0$, $-3y = 15$ so
    $y = -5$ is the y-intercept.
    When $y = 0$, $5x = 15$ so
    $x = 3$ is the x-intercept.

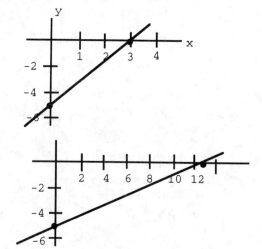

57. When $x = 0$, $-5y = 25$ so $y = -5$
    is the y-intercept.
    When $y = 0$, $2x = 25$, so $x = 12.5$,
    is the x-intercept.

59. Line through $(8, 2)$ and $(3, -3)$ has slope $m_1 = \dfrac{-3 - 2}{3 - 8} = \dfrac{-5}{-5} = 1$

    Line through $(6, -1)$ and $(16, 9)$ has slope $m_2 = \dfrac{9 + 1}{16 - 6} = \dfrac{10}{10} = 1$

       The lines are parallel.

61. Line through $(5, 4)$ and $(1, -2)$ has slope $m_1 = \dfrac{-2 - 4}{1 - 5} = \dfrac{-6}{-4} = \dfrac{3}{2}$

    Line through $(1, 2)$ and $(6, 8)$ has slope $m_2 = \dfrac{8 - 2}{6 - 1} = \dfrac{6}{5}$

    The lines are not parallel.

63. $m_1 = 6 = m_2$ (parallel)

65. The first line may be written $y = \frac{1}{2}x - \frac{3}{2}$ so $m_1 = \frac{1}{2}$.
    The second line may be written $y = -2x + 1$ so $m_2 = -2$.
    The lines are not parallel.

67. $m = 3$ so
    $y - 5 = 3(x + 1)$
    $y = 3x + 3 + 5$
    $y = 3x + 8$

69. $m = -\frac{5}{7}$, $b = 8$
    $y = -\frac{5}{7}x + 8$

71. For Exercise 68 $y = -\frac{3}{2}x + 9$ is written $3x + 2y = 18$

    For Exercise 69 $y = -\frac{5}{7}x + 8$ is written $5x + 7y = 56$

    For Exercise 70 $y = \frac{5}{2}x - 13$ is written as $5x - 2y = 26$

    The coefficients of x and y are the same for a line and a line parallel.

73. $y - 5 = \frac{2}{3}(x - 2)$ When $x = 0$, $y = \frac{2}{3}(0 - 2) + 5 = -\frac{4}{3} + 5 = \frac{11}{3}$, so

    the y-intercept is $\frac{11}{3}$.

75. Let x = no. of weeks from start of the diet and y = weight. Then
    $m = -3$, and (14, 196) is a point on the line.
    (a) $y - 196 = -3(x - 14)$
         $y = -3x + 42 + 196$
         $y = -3x + 238$
    (b) Raul started the diet at $x = 0$ so $y = -3(0) + 238 = 238$ pounds.

77. Let x = number of items and y = the cost. Then (500, 1340) and
    (800, 1760) are points on the line, so
    $m = \frac{1760 - 1340}{800 - 500} = \frac{420}{300} = 1.4$
    $y - 1340 = 1.4(x - 500)$
    $y = 1.4x - 700 + 1340$
    $y = 1.4x + 640$

79. (a) $\frac{y - 3}{-1 - 2} = 3$ (b) $\frac{3 - 2}{x - 1} = -4$
          $y - 3 = -9$ $-4x + 4 = 1$
          $y = -6$ $-4x = -3$ gives $x = \frac{3}{4}$

     (c) $\frac{y - 0}{5 + 2} = \frac{3}{4}$ (d) $\frac{4 + 3}{x + 1} = -\frac{1}{2}$
          $y = 7(\frac{3}{4})$ $-\frac{1}{2}x - \frac{1}{2} = 7$
          $y = \frac{21}{4}$ $-\frac{1}{2}x = \frac{15}{2}$ gives $x = -15$

81. A y-intercept of 9 indicates that (0, 9) is a point on the line.
    Then $m = \frac{9 - 4}{0 - 6} = \frac{5}{-6} = -\frac{5}{6}$ and $y = -\frac{5}{6}x + 9$.

83.　Let x = the number of years since 1980 and y = value of the coin.
Then (0, 185) and (10, 220) are points on the line, so

$$m = \frac{220 - 185}{10 - 0} = \frac{35}{10} = 3.5$$

y - 220 = 3.5(x - 10)
y = 3.5x - 35 + 220
y = 3.5x + 185 is the value x years after 1980.

85.　x = KWH used, y = amount of bill
y = 0.078x + 5

87.　(a) Increases 4　　　　　　　　　　(b)　Decreases 3
　　(c) Increases 2/3　　　　　　　　　(d)　Decreases 1/2

　　(e) $y = -\frac{2}{3}x + \frac{4}{3}$ so it decreases $\frac{2}{3}$　　(f)　No change

89.　Let x = number of miles and y = cost. Then the points (125, 35.75)
　　　　and (265, 51.15) are on the line, so

$$m = \frac{51.15 - 35.75}{265 - 125} = \frac{15.40}{140} = .11$$

y - 35.75 = 0.11(x - 125)
y = 0.11x - 13.75 + 35.75
y = 0.11x + 22

91.　Let x = number of years since 1992 and y = number of cars.
Then m = -15,000 and the y-intercept = 520,000.
y = - 15,000x + 520,000 x years after 1992.

93.　m = 0.28 and the point (24000, 3600) is on the line so
y - 3600 = .28(x - 24000)
y = .28(x - 24,000) + 3600 or y = .28x - 3120

95.　(a)　Let x = number of years with x = 0 for 1980.
　　　　　Let y = birth rate
　　　　　We are given two points (0, 13.7) and (10, 9.9). The slope of
　　　　　the line through these points is -.38 and the y-intercept is
　　　　　13.7 so the linear function is
　　　　　　　　y = -.38x + 13.7   x years after 1980.
　　(b)　For 1985 x = 5 so the birth rate for 1985 is estimated to be
　　　　　　　　y = -.38(5) + 13.7 = 11.8
　　　　　11.8 compares favorably with the actual value of 11.9.
　　(c)　The birth rate will reach zero when y = 0 so
　　　　　　　　0 = -.38x + 13.7
　　　　　　　　x = 13.7/.38 = 36.05
　　　　　This function estimates that Japan's birth rate will drop to
　　　　　zero in the year 1980 + 36 = 2016. This conclusion is based on
　　　　　the assumption that birth rates will drop in a linear manner at
　　　　　the same rate they dropped in 1980-1990. It is unrealistic to
　　　　　expect that no babies will be born in an entire year.

97.　(a)　Let x be the admission price and y be the estimated attendance.
　　　　　The given information provides two points on a line, (5, 185)
　　　　　and (6, 140). The slope of the line through these points is -45
　　　　　and the equation of the line is y - 185 = -45(x - 5) which
　　　　　reduces to y = -45x + 410.
　　(b)　When admission is $7, x = 7 and attendance = -45(7) + 410 = 95.

(c)   When attendance is 250
$$250 = -45x + 410$$
$$x = 3.555$$
For an estimated attendance of 250, the manager would likely round the admission of 3.555 to $3.55 or $3.50.

(d)   For an attendance of zero
$$0 = -45x + 410 \text{ so } x = 9.111$$
An admission of $9.11, or more, would result in no attendance.

(e)   If admission were free, x = 0 and the estimated attendance would be
$$y = -45(0) + 410 = 410$$

103.   About 4:30 am

105.   All have a common y-intercept of 4.

107.   All go through the origin.

109.   (a) 2   (b) 3.5   (c) -8   (d) 0

Section 1.3

1.   (a)   C(180) = 43(180) + 2300 = $10,040
    (b)   Solve 43x + 2300 = 11,889
        43x = 9589   x = 223 bikes
    (c)   Unit cost is $43, fixed cost is $2,300

3.   (a)   Fixed cost is $400, unit cost is $3
    (b)   For 600 units, C(600) = 3(600) + 400 = $2,200
        For 1,000 units, C(1000) = 3(1000) + 400 = $3,400

5.   (a)   R(x) = 32x   (b)   R(78) = 32(78) = $2,496
    (c)   Solve 32x = 672   x = 21 pairs

7.   (a)   R(x) = 3.39x   (b)   R(834) = 3.39(834) = $2,827.26

9.   (a)   C(x) = 57x + 780   (b)   R(x) = 79x
    (c)   79x = 57x + 780   22x = 780
        x = 35.45, so the break-even number is 36 coats.

11.   C(x) = 4x + 500   C(800) = 4(800) + 500 = $3,700

13.   (a)   Let x = number of T-shirts and C = the cost. Then the points (600, 1400) and (700, 1600) lie on the line, so
$$m = \frac{1600 - 1400}{700 - 600} = \frac{200}{100} = 2$$
$$y - 1600 = 2(x - 700)$$
$$y = 2x - 1400 + 1600$$
$$C(x) = 2x + 200$$
    (b) $200   (c) $2

15.   (a)   C(x) = 649x + 1500   (b)   R(x) = 899x
    (c)   C(37) = 649(37) + 1500 = $25,513
    (d)   R(37) = 899(37) = $33,263
    (e)   899x = 649x + 1500
        250x = 1500   x = 6 computers

17.   (a)   Let x = number of years and BV = the book value.
Then the points (0, 425) and (8, 25) are on the line, so
$$m = \frac{425 - 25}{0 - 8} = \frac{400}{-8} = -50$$
BV - 425 = -50x
BV = -50x + 425

(b)   Annual depreciation is $50

(c)   BV(3) = -50(3) + 425 = -150 + 425 = $275

19.   Let x = number of years and BV = the book value. Then the points
(0, 9750) and (6, 300) lie on the line, so
$$m = \frac{9750 - 300}{0 - 6} = \frac{9450}{-6} = -1575 \text{ and the y-intercept} = 9750.$$

(a)   BV = -1575x + 9750

(b) $1,575

(c)   BV(2) = -1575(2) + 9750 = $6600
BV(5) = -1575(5) + 9750 = $1875

21.   Revenue must be greater than costs, so
0.45x > 0.23x + 475
0.22x > 475
x > 2159.09 at break-even, so at least 2,160 cookies must be sold to make a profit.

23.   Let x = number of miles traveled. The weekly costs for Company A,
C(x) = .14x + 105, must be less than the weekly costs for Company B,
C(x) = .10x + 161.
0.14x + 105 < 0.10x + 161
0.04x < 56
x < 1400
Company A is better when the weekly mileage is less than 1400 miles.

25.   Let x = number of copies sold. A profit occurs when
.85x + .20x > .40x + 1400
.65x > 1400
x > 2153.846, so at least 2,154 copies must be sold to make a profit.

27.   Let x = number of books. The unit cost gives m = 12.65. The point
(2700, 36295) lies on the line so
y - 36295 = 12.65(x - 2700)
y = 12.65x + 2140

29.   Let x = number of tickets.

(a)   6x = 650 + 45 + 2.20x
3.8x = 695
x = 182.89, so 183 tickets must be sold to break even.

(b)   6x = 650 + 45 + 2.20x + 700
3.8x = 1395
x = 367.105, so 368 tickets must be sold to clear $700.

(c)   7.5x = 2.20x + 1395
5.3x = 1395
x = 263.208, so 264 tickets must be sold to clear $700.

31.   Let x = number of members

(a)   R(x) = 35x   (b)   R(1238) = 35(1238) = $43,330

(c)   Solve 35x = 595 and obtain x = 17

33.   (a)   Let x = number of years and y = book value, so the points
            (3, 14175) and (7, 8475) lie on the line, so

$$m = \frac{8475 - 14175}{7 - 3} = \frac{-5700}{4} = -1425$$

            y - 14175 = -1425(x - 3)
            y = -1425x + 18,450

    (b) $1,425   (c)   when x = 0, y = $18,450

35.   Let x = number of memberships.
    (a)   Solve a(260) = 3120
            a = 12   so R(x) = 12x
    (b)   Since the break-even membership is 260 and the break-even
          revenue is 3120, the point (260, 3120) is on the cost line.
          The revenue for 200 memberships is $2400 and is $330 less than
          the cost, so the point (200, 2730) is on the cost line. From

$$\text{these two points } m = \frac{3120 - 2730}{260 - 200} = \frac{390}{60} = 6.5$$

         y - 2730 = 6.5(x - 200)
         y = 6.5x - 1300 + 2730
         C(x) = 6.5x + 1430

37.   (a)   Solve a(1465) = 32962.50
            a = 22.5,   so R(x) = 22.5x
    (b)   The points (1465, 26405.50) and (940, 17638) lie on

$$\text{the line, so } m = \frac{26405.50 - 17638}{1465 - 940}$$

$$= \frac{8767.5}{525} = 16.7$$

         y - 17638 = 16.7(x - 940)
         y = 16.7x - 15698 + 17638
         C(x) = 16.7x + 1940
    (c)   Solve 22.5x = 16.7x + 1940
         5.8x = 1940
         x = 334.48 so use 335 for break-even quantity.

39.   Let x be the number of minutes the commercial is aired. Then the cost
    is
              C(x) = 900x + 12,000
    and the revenue is
             R(x) = 2000x
    The break even point occurs when
             2000x = 900x + 12,000
             1100x = 12,000 so   x = 10.9
    The commercial should be aired 11 minutes to recover costs.

41. Let x be the number of knives produced daily. The cost function for the current process is C(x) = 3.85x + 1400. The cost function for the proposed process is
    CP(x) = 2.70x + 1725.
    Graph both functions.

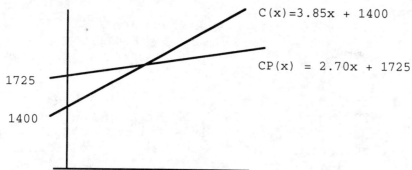

The lines intersect when
    3.85x + 1400 = 2.70x + 1725
    1.15x = 325 gives   x = 282.6
    Since the graph of CP(x) lies below the graph of C(x) when x > 282.6, the new process is more economical when the plant produces more than 282 knives per day.

43. (b) Shortage   (c) Surplus
    (d) When the demand is 20, the price is -$40 so people must be paid to purchase. It is unrealistic to expect a demand of 20.

45. The maximum production level is 595

47. Let x = monthly sales. Then the three plans are
    Plan 1: y = 2500
    Plan 2: y = .17x + 2000
    Plan 3: y = .25x + 1700

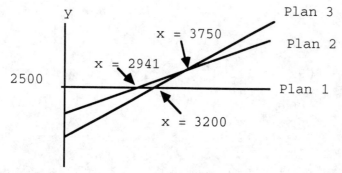

Plan 1 and Plan 2 intersect when 2500 = 2000 + .17x, when x = $2,941.
Plan 1 and Plan 3 intersect when 2500 = 1700 + .25x, when x = $3,200.
Plan 2 and Plan 3 intersect when 2000 + .17x = 1700 + .25x, when x = $3,750.
Plan 1 is better when sales are less than $2941.
Plan 2 is better when sales are between $2941 and $3750.
Plan 3 is better when sales are greater than $3750.

49. (a) Let x = number of tickets sold. The cost at the Convention Center is C(x) = 20x + 600. The cost at the Ferrell Center is C(x) = 17x + 1300. Break even at the Convention Center occurs when 35x = 20x + 600 + 2000 + 700 = 20x + 3300. x = 220.

10

(b) Break even at the Ferrell Center occurs when 35x = 17x + 1300 + 2000 + 700 = 17x + 4000. x = 222.22. They must sell 223 to break even.

(c) At the Convention Center the profit is
P(x) = 35x - 20x - 3300 = 15x -3300
At the Ferrell Center the profit is
P(x) = 35x - 17x -4000 = 18x - 4000
These lines intersect at x = 233.3. The Ferrell Center is more profitable when more than 233 tickets are sold, otherwise the Convention Center is more profitable.

51. Let x = number of trips.
Plan 1: y = 2500
Plan 2: y = 130x + 300
Plan 3: y = 165x

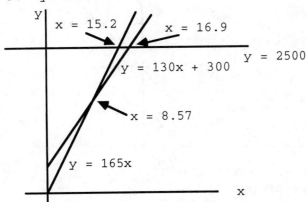

The lines for Plan 1 and 2 intersect when 2500 = 130x + 300, that is, when x = 16.9.
The lines for Plan 1 and 3 intersect when 2500 = 165x, that is, when x = 15.2.
The lines for Plan 2 and 3 intersect when 130x + 300 = 165x, that is, when x = 8.57.
Plan 1 is better for 8 trips or less, Plan 2 is better for 9 through 16 trips, and Plan 3 is better for more than 16 trips.

Review Exercises, Chapter 1

1. (a) $f(5) = \dfrac{7 \times 5 - 3}{2} = 16$ (b) $f(1) = 2$

    (c) $f(4) = 12.5$ (d) $f(b) = \dfrac{7b - 3}{2}$

3. $f(2) + g(3) = \dfrac{2 + 2}{2 - 1} + 5(3) + 3 = \dfrac{4}{1} + 15 + 3 = 22$

5. (a) $f(3.5) = 1.20(3.5) = \$4.20$
    (b) Solve 1.20x = 3.30
         x = 2.75 pounds

7. (a) $f(x) = 29.95x$ (b) $f(x) = 1.25x + 40$

9.   (a)   (b)

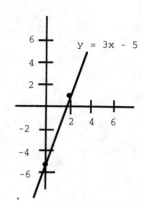   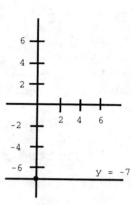

        (c)   (d)

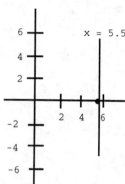   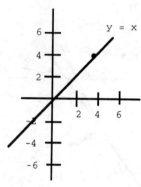

11.   (a)   Slope is -2, y-intercept is 3
        (b)   Slope is 2/3, y-intercept is -4

        (c) $y = \frac{5}{4}x + \frac{3}{2}$ slope is $\frac{5}{4}$, y-intercept is $\frac{3}{2}$

        (d) $y = -\frac{6}{7}x - \frac{5}{7}$ slope is $-\frac{6}{7}$, y-intercept is $-\frac{5}{7}$

13. $y = -\frac{6}{5}x + 3$ (a) $m = -\frac{6}{5}$ (b) y-intercept = 3
        (c)   When y = 0, x = 5/2, so x-intercept = 5/2

15.   (a) $y = -\frac{3}{4}x + 5$ (b)   y = 8x - 3

        (c) $y + 1 = -2(x - 5)$ (d)   Horizontal line, y = 6
                $y = -2x + 10 - 1$
                $y = -2x + 9$

        (e) $m = \frac{4 - 3}{-1 - 5} = \frac{1}{-6}$ (f)   Vertical line, x = -2

                $y - 3 = -\frac{1}{6}(x - 5)$

                $y = -\frac{1}{6}x + \frac{5}{6} + 3$

                $y = -\frac{1}{6}x + \frac{23}{6}$

        (g) $y = \frac{4}{3}x - \frac{22}{3}$, so $m = \frac{4}{3}$

                $y - 7 = \frac{4}{3}(x - 2)$

                $y = \frac{4}{3}x + \frac{13}{3}$ or 4x - 3y = -13

12

17. (a) $m = \dfrac{2 - 2}{-3 - 6} = 0$ (b) $m = \dfrac{-2 - 5}{-4 + 4}$

$y - 2 = 0(x - 6)$ Slope undefined, $x = -4$

$y = 2$

(c) $m = \dfrac{10 - 0}{5 - 5}$ Slope undefined, $x = 5$

(d) $m = \dfrac{6 - 6}{7 + 7} = 0$ $y = 6$

19. The line through (5, 19) and (-2, 7) has slope $m_1 = \dfrac{19 - 7}{5 + 2} = \dfrac{12}{7}$

The line through (11, 3) and (-1, -5) has slope

$m_2 = \dfrac{3 + 5}{11 + 1} = \dfrac{8}{12} = \dfrac{2}{3}$ The lines are not parallel.

21. The given line has slope $m = -2$. The line through (8, 6) and

(-3, 14) has slope $m = \dfrac{14 - 6}{-3 - 8} = \dfrac{8}{-11} = -\dfrac{8}{11}$

The lines are not parallel.

23. The given line has slope $m = \dfrac{3}{2}$. The line through (9, 10) and

(5, 6) has slope $m = \dfrac{10 - 6}{9 - 5} = 1$ The lines are not parallel.

25. Let x = number of items produced.
$C(x) = 36x + 12,800$

27. (a) $C(580) = 3.60(580) + 2850 = \$4,938$
(b)   Solve $3.60x + 2850 = 5208$
$3.60x = 2358$ which gives $x = 655$ bags

29. (a) $R(x) = 11x$ (b) $C(x) = 6.5x + 675$
(c)   Solve $11x = 6.5x + 675$
$4.5x = 675$ giving $x = 150$

31. $R(x) = 19.5x$
The points (1840, 25260) and (2315, 31102.5) lie on the cost line
so, $m = \dfrac{31102.5 - 25260}{2315 - 1840} = \dfrac{5842.5}{475} = 12.3$

$y - 25260 = 12.3(x - 1840)$
$C(x) = 12.3x + 2628$
To find the break-even quantity, solve
$19.5x = 12.3x + 2628$
$7.2x = 2628$ gives $x = 365$ watches

33. (a)   The points (0, 17500) and (8, 900) lie on the line
so $m = \dfrac{900 - 17500}{8 - 0} = \dfrac{-16600}{8} = -2075$

$BV = -2075x + 17500$
(b) $\$2,075$
(c) $BV(5) = -2075(5) + 17500 = \$7,125$

35. The points (0, 1540) and (5, 60) lie on the line, so
$m = \dfrac{1540 - 60}{0 - 5} = \dfrac{1480}{-5} = -296$

$BV = -296x + 1540$

37. The slope of the given line is 4/5, so $\dfrac{k - 9}{-3 - 2} = \dfrac{4}{5}$ which gives
$k - 9 = -4$ and $k = 5$

39. The points (0, 12000) and (5, 1500) lie on the line, so

$$m = \frac{1500 - 12000}{5 - 0} = \frac{-10500}{5} = -2100$$

BV = -2100x + 12,000

41. Let x = number of hamburgers and y = cost. Then m = 0.67, and the
point (1150, 1250.5) lies on the line,
so y - 1250.5 = 0.67(x - 1150)
y = 0.67x - 770.5 + 1250.5
C(x) = 0.67x + 480

43. Let x = number of items sold. The second option is better when
0.75x + 2000 > 17000
0.75x > 15000   x > 20,000
The second plan is better when sales exceed 20,000 items.

45. $$\frac{k - 4}{-2 - 9} = -2$$

k - 4 = 22 so k = 26

47. Let x = number of tapes and C(x) = total cost. Since the unit cost
is $6.82, m = 6.82, and the point (1730, 12813.60) lies on the line,
y - 12813.60 = 6.82(x - 1730)
C(x) = 6.82x + 1015
The fixed cost is $1015.

49.

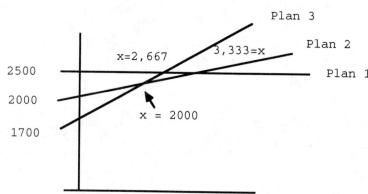

Let x = sales, then the plans can be represented by
Plan 1: y = 2500
Plan 2: y = .15x + 2000
Plan 3: y = .30x + 1700
Plan 1 and Plan 2 intersect at 2500 = 2000 + .15x, x = 3,333.
Plan 1 and Plan 3 intersect at 2500 = 1700 + .30x, x = 2,667.
Plan 2 and Plan 3 intersect at 2000 + .15x = 1700 + .30x, x = 2000.
Plan 1 is better when sales are less than $2,667.
Plan 2 is never better.
Plan 3 is better when sales are larger than $2,667.

51. (a)   y = 120x + 8400
     (b)   10,000 = 120x + 8400
                    x = 13.33
        The holdings will reach 10,000 books during the 14th month.

53. y = 500x - 1700

Section 2.1

1. $4x - y = 5$
    $x + 2y = 8$

    $x = 8 - 2y$
    $4(8 - 2y) - y = 5$
    $32 - 8y - y = 5$
    $\qquad -9y = -27$
    $\qquad\quad y = 3$
    $x = 8 - 2(3) = 2$ $\qquad\qquad$ $(2, 3)$

3. $5x - y = -15$
    $x + y = -3$

    $x = -3 - y$
    $5(-3 - y) - y = -15$
    $\qquad -15 - 5y - y = -15$
    $\qquad\qquad -6y = 0$
    $\qquad\qquad\quad y = 0$
    $x = -3 - 0 = -3$ $\qquad\qquad$ $(-3, 0)$

5. $y = 5x$
    $6x - 2y = 12$

    $\qquad 6x - 2(5x) = 12$
    $\qquad\qquad -4x = 12$
    $\qquad\qquad\quad x = -3$
    $y = 5(-3) = -15$ $\qquad\qquad$ $(-3, -15)$

7. $7x - y = 32$
    $2x + 3y = 19$

    $y = 7x - 32$
    $2x + 3(7x - 32) = 19$
    $\qquad\quad 2x + 21x - 96 = 19$
    $\qquad\qquad 23x = 115$
    $\qquad\qquad\quad x = 5$
    $y = 7(5) - 32 = 3$ $\qquad\qquad$ $(5, 3)$

9. $5x + 2y = 14$
    $x - 3y = 30$

    $x = 3y + 30$
    $5(3y + 30) + 2y = 14$
    $\qquad 15y + 150 + 2y = 14$
    $\qquad\qquad 17y = -136$
    $\qquad\qquad\quad y = -8$
    $x = 3(-8) + 30 = 6$ $\qquad\qquad$ $(6, -8)$

11. $22x + y = 81$
    $8x - 3y = 16$

    $y = 81 - 22x$
    $8x - 3(81 - 22x) = 16$
    $\qquad 8x - 243 + 66x = 16$
    $\qquad\quad 74x = 259$
    $\qquad\qquad x = 3.5$
    $y = 81 - 22(3.5) = 4$ $\qquad\qquad$ $(3.5, 4)$

13. $6x - 3y = 9$
    $9x - 15y = 31$

    $3y = 6x - 9$
    $y = 2x - 3$
    $\qquad 9x - 15(2x - 3) = 31$
    $\qquad\quad 9x - 30x + 45 = 31$
    $\qquad -21x = -14$ so $x = 2/3$
    $y = 2(2/3) - 3 = 4/3 - 3 = -5/3$
    $\qquad\qquad\qquad\qquad$ $(2/3, -5/3)$

15. $y = 3x - 5$
    $8x - 4y - 30 = 0$

    $8x - 4(3x - 5) - 30 = 0$
    $\qquad 8x - 12x + 20 - 30 = 0$
    $\qquad -4x - 10 = 0$ so $x = -2.5$
    $y = -12.5$ $\qquad\qquad$ $(-2.5, -12.5)$

17. $3x - 4y = 22$
     $2x + 5y = 7$

Multiply first equation by 2 and the second by -3 to eliminate x.

$$6x - 8y = 44$$
$$\underline{-6x - 15y = -21}$$
$$-23y = 23$$
$$y = -1$$
$$3x - 4(-1) = 22$$
$$3x + 4 = 22$$
$$x = 6 \qquad\qquad (6, -1)$$

19. $6x - y = 18$
     $2x + y = 2$

$$6x - y = 18$$
$$\underline{2x + y = 2}$$
$$8x = 20$$
$$x = 5/2$$
$$2(5/2) + y = 2$$
$$5 + y = 2$$
$$y = -3 \qquad\qquad (5/2, -3)$$

21. $-2x + y = 7$
     $6x + 12y = 24$

Multiply first equation by 3 to eliminate x.

$$-6x + 3y = 21$$
$$\underline{6x + 12y = 24}$$
$$15y = 45$$
$$y = 3$$
$$-2x + 3 = 7$$
$$-2x = 4$$
$$x = -2 \qquad\qquad (-2, 3)$$

23. $2x + y = -9$
     $4x + 3y = 1$

Multiply first equation by -2 to eliminate x

$$-4x - 2y = 18$$
$$\underline{4x + 3y = 1}$$
$$y = 19$$
$$2x + 19 = -9$$
$$2x = -28$$
$$x = -14 \qquad\qquad (-14, 19)$$

25. $7x + 3y = -1.5$
     $2x - 5y = -30.3$

Multiply first equation by 5 and the second by 3 to eliminate x.

$$35x + 15y = -7.5$$
$$\underline{6x - 15y = -90.9}$$
$$41x = -98.4$$
$$x = -2.4$$
$$2(-2.4) - 5y = -30.3$$
$$-5y = -25.5$$
$$y = 5.1 \qquad\qquad (-2.4, 5.1)$$

27. $2x - 3y = -0.27$
     $5x - 2y = 0.04$

Multiply first equation by -2 and the second by 5 to eliminate y.

$$-4x + 6y = 0.54$$
$$\underline{15x - 6y = 0.12}$$
$$11x = 0.66$$
$$x = 0.06$$
$$2(0.06) - 3y = -0.27$$
$$-3y = -0.39$$
$$y = 0.13 \qquad\qquad (0.06, 0.13)$$

29. $6x - 9y = 8$
     $10x - 15y = -20$

Multiply first equation by 5 and the second by -3 to eliminate x.
$$30x - 45y = 40$$
$$\underline{-30x + 45y = 60}$$
$$0 = 100$$
No solution

31. $8x + 10y = 2$
     $12x + 15y = 3$

Divide first equation by 2 and the second by -3 to eliminate x.
$$4x + 5y = 1$$
$$\underline{-4x - 5y = -1}$$
$$0 = 0$$
Infinite number of solutions

33. $x - 6y = 4$
     $5x - 30y = 20$

Divide second equation by -5
$$x - 6y = 4$$
$$\underline{-x + 6y = -4}$$
$$0 = 0$$
Infinite number of solutions

35. $p = -3x + 15$
     $p = 2x - 5$

$2x - 5 = -3x + 15$
$5x = 20$
$x = 4$
$p = -3(4) + 15 = 3$ \qquad (4, 3)
Equilibrium when x = 4 and p = 3

37. $p = -4x + 130$
     $p = x - 20$

$x - 20 = -4x + 130$
$5x = 150$
$x = 30$
$p = 30 - 20 = 10$ \qquad (30, 10)
Equilibrium when x = 30 and p = 10

39. $p = -5x + 83$
     $p = 4x - 52$

$4x - 52 = -5x + 83$
$9x = 135$
$x = 15$
$p = 4(15) - 52 = 8$ \qquad (15, 8)
Equilibrium when x = 15 and p = 8

41. $p = -2.5x + 148$
     $p = 1.7x + 43$

$1.7x + 43 = -2.5x + 148$
$4.2x = 105$
$x = 25$
$p = 1.7(25) + 43 = 85.5$ \qquad (25, 85.5)

43. Let x = number of oranges and y = number of apples.
     $50x + 8y = 151$ (Calcium)
     $0.5x + 0.4y = 2.55$ (Iron)

$$50x + 8y = 151$$
$$\underline{-50x - 40y = -255}$$
$$-32y = -104$$
$y = 3.25$
$50x + 8(3.25) = 151$
$50x = 125$
$x = 2.5$
2.5 oranges, 3.25 apples

45. (a) $C(x) = .85x + 160,000$, $R(x) = 5x$
The break-even point occurs when
$5x = .85x + 160,000$
$4.15x = 160,000$
$x = 38,554$
They must sell 38,554 CD's in order to break even.
(b)   If 20,000 CD's are sold, the costs are
$.85(20,000) + 160,000 = 177,000$. The revenue is $5(20,000) = 100,000$. Thus, a loss of $77,000 occurs.
(c)   If 50,000 CD's are sold, the costs are
$.85(50,000) + 160,000 = 202,500$ and the revenue is $5(50,000) = 250,000$. Thus, a profit of $47,500 occurs.

47. Let n = number of nickels and d = number of dimes.
$n + d = 165$ $d = 165 - n$
$5n + 10d = 1435$ $5n + 10(165 - n) = 1435$
$5n + 1650 - 10n = 1435$
$-5n = -215$
$n = 43$
$d = 165 - 43 = 122$
43 nickels, 122 dimes

49. Let x = gallons of 20% acid and y = gallons of 55% acid. The solution is to contain 30% of 840 gallons, which is 252 gallons.
$x + y = 840$
$.20x + .55y = .30(840)$ Multiply the second equation by 100
and substitute $x = 840 - y$
$20(840 - y) + 55y = 25200$
$35y = 8400$
$y = 240$
$x = 840 - 240 = 600$
600 gallons of 20%, 240 gallons of 55%

51. Let x = number produced at McGregor and y = number produced at Ennis.
$x + y = 1500$
At McGregor $C(x) = 8.40x + 7480$
At Ennis $C(x) = 7.80y + 5419$
If costs are equal, then
$8.4x + 7480 = 7.8y + 5419$
$x = 1500 - y$
$84(1500 - y) - 78y = -20610$
$162y = 146610$
$y = 905$ so $x = 1500 - 905 = 595$
595 at McGregor, 905 at Ennis

53. Let P = weight of peanuts and C = weight of cashews.
$P + C = 80$ $P = 80 - C$
$3P + 7C = 4(80)$ $3(80 - C) + 7C = 320$
$4C = 80$
$C = 20$
$P = 80 - 20 = 60$
60 lb of peanuts, 20 lb of cashews
587 tie tacks, 213 lapel pins

55.   Let x = cases of Golden Punch and y = cases of Light Punch.
        4x + 7y = 142   (Apple)   12x + 21y = 426
        6x + 3y = 108   (Pineapple)   -12x - 6y = -216
                                              15y = 210
                                                y = 14
                                                x = 11
                     11 cases of Golden, 14 cases of Light Punch

57.   Let B = number of beef and S = number of sausage sandwiches.
           B +   S = 115   S = 115 - B
        2.3B + 2.1S = 253.90   23B + 21(115 - B) = 2539
                                          2B = 124
                                            B = 62 so   S = 115 - 62 = 53
                             62 chopped beef, 53 sausage

59.   Let x = amount in tax-free and y = amount in money market.
            x + y = 50000   x = 50000 - y
        .074x + .088y = 4071   74(50000 - y) + 88y = 4071000
                                        14y = 371,000
                                          y = 26,500
                                          x = 23,500
                    $23,500 in tax-free, $26,500 in money market

61.   Let x = amount of federal tax and y = amount of state tax.
        Federal tax = x = .20(198000 - y)
        State tax = y = .05(198000 - x)
        These equations reduce to
              x + .2y = 39600   .05x + .01y = 1980
            .05x +   y = 9900   -.05x -   y = -9900
                                              -.99y = -7920
                                                  y = 8,000
                                        x = .2(190,000)= 38,000
                    Federal tax is $38,000, state tax is $8,000

63.   Let x = number of years later and y = total deposits.
        For Central Bank y = 250,000x + 3,000,000.
        For Citizens Bank y = -300,000x + 9,000,000.
              Their deposits are equal when 250,000x + 3,000,000 =
        -300,000x + 9,000,000.
              550,000x = 6,000,000 so   x = 10.91
        Their deposits will be equal in a little less than 11 years.

65.   (a)   x = 4.54, y = 7.74   (b)   x = .75, y = 9.25
        (c)   x = 1.26, y = 4.80   (d)   x = 3.02, y = -.72

67.   20.55 tricycles with p = $117.53. If we round to 21 tricycles,
        the demand price is $118.80 and the supply price is $115.50

69.   (a)   Equilibrium at (27, 126)
        (b)   When p = $110, x = 11 in the supply equation and x = 35 in
              the demand equation so there is a shortage of 24.
        (c)   When p = $135, x = 36 in the supply equation and x = 22.5 in
              the demand equation so there is a surplus of 13.5.

Section 2.2

1.
$$x + 2y = 7$$
$$3x + 5y = 19$$

$$3x + 6y = 21$$
$$\underline{3x + 5y = 19}$$
$$y = 2$$
$$x + 2(2) = 7$$
$$x = 3$$

(3, 2)

3.
$$2x + 5y = -1$$
$$6x - 4y = 16$$

$$6x + 15y = -3$$
$$\underline{6x - 4y = 16}$$
$$19y = -19$$
$$y = -1$$
$$2x + 5(-1) = -1$$
$$x = 2 \qquad (2, -1)$$

5.
$$x + y - z = -1$$
$$x - y + z = 5$$
$$x - y - z = 1$$

$$x + y - z = -1 \qquad \text{subtract second}$$
$$\underline{x - y + z = 5} \qquad \text{from first}$$
$$2y - 2z = -6$$
$$x - y + z = 5 \qquad \text{subtract third}$$
$$\underline{x - y - z = 1} \qquad \text{from second}$$
$$2z = 4$$
$$z = 2$$
Substitute z = 2
$$2y - 2(2) = -6$$
$$2y = -2$$
$$y = -1$$
Substitute y = -1 and z = 2
$$x + (-1) - 2 = -1$$
$$x = 2$$
(2, -1, 2)

7.
$$x + 4y - 2z = 21$$
$$3x - 6y - 3z = -18$$
$$2x + 4y + z = 37$$

$$3x + 12y - 6z = 63 \qquad \text{multiply first equation by 3}$$
$$\underline{3x - 6y - 3z = -18} \qquad \text{subtract second}$$
$$18y - 3z = 81$$
$$2x + 8y - 4z = 42 \qquad \text{multiply first equation by 2}$$
$$\underline{2x + 4y + z = 37} \qquad \text{subtract third}$$
$$4y - 5z = 5$$
Use these two new equations
$$18y - 3z = 81$$
$$4y - 5z = 5$$

$$36y - 6z = 162$$
$$\underline{36y - 45z = 45}$$
$$39z = 117$$
$$z = 3$$
Substitute z = 3
$$18y - 3(3) = 81$$
$$18y = 90$$
$$y = 5$$
Substitute z = 3 and y = 5
$$x + 4(5) - 2(3) = 21$$
$$x = 7 \qquad\qquad (7, 5, 3)$$

9. $2x + 4y - 6z = -2$
    $4x - 3y + z = 11$
    $3x + 2y - 2z = 7$

    $4x + 8y - 12z = -4$ multiply first equation by 2
    $\underline{4x - 3y + z = 11}$ subtract second
    $11y - 13z = -15$
    $6x + 12y - 18z = -6$ multiply first equation by 3
    $\underline{6x + 4y - 4z = 14}$ multiply third equation by 2
    $8y - 14z = -20$ subtract
    Now solve $\quad 11y - 13z = -15$
    $\qquad\qquad\qquad 8y - 14z = -20$
    $88y - 104z = -120$
    $\underline{88y - 154z = -220}$
    $50z = 100$
    $z = 2$
    $11y - 13(2) = -15$
    $11y = 11$
    $y = 1$
    $2x + 4(1) - 6(2) = -2$
    $2x = 6$
    $x = 3$
    $(3, 1, 2)$

11. (a) $\quad a_{11} = 2 \qquad a_{22} = 3 \qquad a_{33} = 6 \qquad a_{43} = 11$
    (b) $\quad (2, 3)$
    (c) $\quad a_{12} = 4 \qquad a_{32} = 0 \qquad a_{41} = 9$

13. coeff: $\begin{bmatrix} 5 & -2 \\ 3 & 1 \end{bmatrix}$ aug: $\left[\begin{array}{cc|c} 5 & -2 & 1 \\ 3 & 1 & 7 \end{array}\right]$

15. coeff: $\begin{bmatrix} 1 & 1 & -1 \\ 3 & 4 & -2 \\ 2 & 0 & 1 \end{bmatrix}$ aug: $\left[\begin{array}{ccc|c} 1 & 1 & -1 & 14 \\ 3 & 4 & -2 & 9 \\ 2 & 0 & 1 & 7 \end{array}\right]$

17. coeff: $\begin{bmatrix} 1 & 5 & -2 & 1 \\ 1 & -1 & 2 & 4 \\ 6 & 3 & -11 & 1 \\ 5 & -3 & -7 & 1 \end{bmatrix}$ aug: $\left[\begin{array}{cccc|c} 1 & 5 & -2 & 1 & 12 \\ 1 & -1 & 2 & 4 & -5 \\ 6 & 3 & -11 & 1 & 14 \\ 5 & -3 & -7 & 1 & 22 \end{array}\right]$

19. $5x + 3y = -2$
    $-x + 4y = 4$

21. $5x_1 + 2x_2 - x_3 = 3$
    $-2x_1 + 7x_2 + 8x_3 = 7$
    $3x_1 \qquad\quad + x_3 = 5$

23. $3x_1 \qquad\quad + 2x_3 + 6x_4 = 4$
    $-4x_1 + 5x_2 + 7x_3 + 2x_4 = 2$
    $x_1 + 3x_2 + 2x_3 + 5x_4 = 0$
    $-2x_1 + 6x_2 - 5x_3 + 3x_4 = 4$

25. $\left[\begin{array}{ccc|c} 1 & 2 & -4 & 6 \\ 4 & 2 & 5 & 7 \\ 1 & -1 & 0 & 4 \end{array}\right]$

27. $\left[\begin{array}{ccc|c} 1 & 3 & 2 & -4 \\ 0 & -7 & -1 & 13 \\ 0 & -6 & -10 & 19 \end{array}\right]$

29. $\left[\begin{array}{ccc|c} 1 & -3 & 2 & -6 \\ 0 & 1 & -2 & 4 \\ 0 & 4 & 3 & 8 \end{array}\right]$

31. $\begin{bmatrix} 2 & 3 & | & 5 \\ 1 & -2 & | & -1 \end{bmatrix}$ R1 ↔ R2 $\begin{bmatrix} 1 & -2 & | & -1 \\ 2 & 3 & | & 5 \end{bmatrix}$ -2R1 + R2 → R2

$\begin{bmatrix} 1 & -2 & | & -1 \\ 0 & 7 & | & 7 \end{bmatrix}$ (1/7)R2 → R2 $\begin{bmatrix} 1 & -2 & | & -1 \\ 0 & 1 & | & 1 \end{bmatrix}$ 2R2 + R1 → R1

$\begin{bmatrix} 1 & 0 & | & 1 \\ 0 & 1 & | & 1 \end{bmatrix}$ x = 1, y = 1

33. $\begin{bmatrix} 1 & -3 & | & -1 \\ 4 & 5 & | & 30 \end{bmatrix}$ -4R1 + R2 → R2 $\begin{bmatrix} 1 & -3 & | & -1 \\ 0 & 17 & | & 34 \end{bmatrix}$ (1/17)R2 → R2

$\begin{bmatrix} 1 & -3 & | & -1 \\ 0 & 1 & | & 2 \end{bmatrix}$ 3R2 + R1 → R1 $\begin{bmatrix} 1 & 0 & | & 5 \\ 0 & 1 & | & 2 \end{bmatrix}$ x = 5, y = 2

35. $\begin{bmatrix} 2 & 4 & | & -7 \\ 1 & -3 & | & 9 \end{bmatrix}$ R1 ↔ R2 $\begin{bmatrix} 1 & -3 & | & 9 \\ 2 & 4 & | & -7 \end{bmatrix}$ -2R1 + R2 → R2

$\begin{bmatrix} 1 & -3 & | & 9 \\ 0 & 10 & | & -25 \end{bmatrix}$ (1/10)R2 → R2 $\begin{bmatrix} 1 & -3 & | & 9 \\ 0 & 1 & | & -5/2 \end{bmatrix}$ 3R2 + R1 → R1

$\begin{bmatrix} 1 & 0 & | & 3/2 \\ 0 & 1 & | & -5/2 \end{bmatrix}$ x = 3/2, y = -5/2

37. $\begin{bmatrix} 1 & 2 & -1 & | & 3 \\ 1 & 3 & -1 & | & 4 \\ 1 & -1 & 1 & | & 4 \end{bmatrix}$ -R1 + R2 → R2
-R1 + R3 → R3 $\begin{bmatrix} 1 & 2 & -1 & | & 3 \\ 0 & 1 & 0 & | & 1 \\ 0 & -3 & 2 & | & 1 \end{bmatrix}$ -2R1 + R1 → R1
3R2 + R3 → R3

$\begin{bmatrix} 1 & 0 & -1 & | & 1 \\ 0 & 1 & 0 & | & 1 \\ 0 & 0 & 2 & | & 4 \end{bmatrix}$ (1/2)R3 → R3 $\begin{bmatrix} 1 & 0 & -1 & | & 1 \\ 0 & 1 & 0 & | & 1 \\ 0 & 0 & 1 & | & 2 \end{bmatrix}$ R3 + R1 → R1

$\begin{bmatrix} 1 & 0 & 0 & | & 3 \\ 0 & 1 & 0 & | & 1 \\ 0 & 0 & 1 & | & 2 \end{bmatrix}$ $x_1 = 3,\ x_2 = 1,\ x_3 = 2$

39. $\begin{bmatrix} 2 & 4 & 2 & | & 6 \\ 2 & 1 & 1 & | & 16 \\ 1 & 1 & 2 & | & 9 \end{bmatrix}$ R1 ↔ R3 $\begin{bmatrix} 1 & 1 & 2 & | & 9 \\ 2 & 1 & 1 & | & 16 \\ 2 & 4 & 2 & | & 6 \end{bmatrix}$ -2R1 + R2 → R2
-2R1 + R3 → R3

$\begin{bmatrix} 1 & 1 & 2 & | & 9 \\ 0 & -1 & -3 & | & -2 \\ 0 & 2 & -2 & | & -12 \end{bmatrix}$ R2 + R1 → R1
-R2 → R2
2R2 + R3 → R3 $\begin{bmatrix} 1 & 0 & -1 & | & 7 \\ 0 & 1 & 3 & | & 2 \\ 0 & 0 & -8 & | & -16 \end{bmatrix}$ -(1/8)R3 → R3

$\begin{bmatrix} 1 & 0 & -1 & | & 7 \\ 0 & 1 & 3 & | & 2 \\ 0 & 0 & 1 & | & 2 \end{bmatrix}$ R3 + R1 → R1
-3R3 + R2 → R2 $\begin{bmatrix} 1 & 0 & 0 & | & 9 \\ 0 & 1 & 0 & | & -4 \\ 0 & 0 & 1 & | & 2 \end{bmatrix}$ (9, -4, 2)

41. $\begin{bmatrix} 1 & 2 & -1 & | & -1 \\ 2 & -3 & 2 & | & 15 \\ 0 & 1 & 4 & | & -7 \end{bmatrix}$ -2R1 + R2 → R2 $\begin{bmatrix} 1 & 2 & -1 & | & -1 \\ 0 & -7 & 4 & | & 17 \\ 0 & 1 & 4 & | & -7 \end{bmatrix}$ R2 ↔ R3

$\begin{bmatrix} 1 & 2 & -1 & | & -1 \\ 0 & 1 & 4 & | & -7 \\ 0 & -7 & 4 & | & 17 \end{bmatrix}$ -2R2 + R1 → R1
7R2 + R3 → R3 $\begin{bmatrix} 1 & 0 & -9 & | & 13 \\ 0 & 1 & 4 & | & -7 \\ 0 & 0 & 32 & | & -32 \end{bmatrix}$

$\begin{bmatrix} 1 & 0 & -9 & | & 13 \\ 0 & 1 & 4 & | & -7 \\ 0 & 0 & 1 & | & -1 \end{bmatrix}$ $\begin{bmatrix} 1 & 0 & 0 & | & 4 \\ 0 & 1 & 0 & | & -3 \\ 0 & 0 & 1 & | & -1 \end{bmatrix}$ (4, -3, -1)

43. $\begin{bmatrix} 1 & 5 & -1 & | & 1 \\ 4 & -2 & -3 & | & 6 \\ -3 & 1 & 3 & | & -3 \end{bmatrix}$ $\begin{bmatrix} 1 & 5 & -1 & | & 1 \\ 0 & -22 & 1 & | & 2 \\ 0 & 16 & 0 & | & 0 \end{bmatrix}$

$\begin{bmatrix} 1 & 0 & -1 & | & 1 \\ 0 & 0 & 1 & | & 2 \\ 0 & 1 & 0 & | & 0 \end{bmatrix}$ $\begin{bmatrix} 1 & 0 & 0 & | & 3 \\ 0 & 0 & 1 & | & 2 \\ 0 & 1 & 0 & | & 0 \end{bmatrix}$ (3, 0, 2)

45. $\begin{bmatrix} 2 & 2 & 4 & | & 16 \\ 1 & 2 & -1 & | & 6 \\ -3 & 1 & -2 & | & 0 \end{bmatrix}$ $\begin{bmatrix} 1 & 1 & 2 & | & 8 \\ 0 & 1 & -3 & | & -2 \\ 0 & 1 & 1 & | & 6 \end{bmatrix}$

$\begin{bmatrix} 1 & 0 & 5 & | & 10 \\ 0 & 1 & -3 & | & -2 \\ 0 & 0 & 1 & | & 2 \end{bmatrix}$ $\begin{bmatrix} 1 & 0 & 0 & | & 0 \\ 0 & 1 & 0 & | & 4 \\ 0 & 0 & 1 & | & 2 \end{bmatrix}$ (0, 4, 2)

47. $\begin{bmatrix} 1 & 2 & -3 & | & -6 \\ 1 & -3 & -7 & | & 10 \\ 1 & -1 & 1 & | & 10 \end{bmatrix}$ $\begin{bmatrix} 1 & 2 & -3 & | & -6 \\ 0 & 5 & 4 & | & -16 \\ 0 & -3 & 4 & | & 16 \end{bmatrix}$

$\begin{bmatrix} 1 & 0 & -23/5 & | & 2/5 \\ 0 & 1 & 4/5 & | & -16/5 \\ 0 & 0 & 32/5 & | & 32/5 \end{bmatrix}$ $\begin{bmatrix} 1 & 0 & 0 & | & 5 \\ 0 & 1 & 0 & | & -4 \\ 0 & 0 & 1 & | & 1 \end{bmatrix}$ (5, -4, 1)

49. $\begin{bmatrix} 1 & 1 & 1 & 1 & | & 4 \\ 1 & 2 & -1 & -1 & | & 7 \\ 2 & -1 & -1 & -1 & | & 8 \\ 1 & -1 & 2 & -2 & | & -7 \end{bmatrix}$ $\begin{array}{l} -R1 + R2 \to R2 \\ \\ -2R1 + R3 \to R3 \\ -R1 + R4 \to R4 \end{array}$

$\begin{bmatrix} 1 & 1 & 1 & 1 & | & 4 \\ 0 & 1 & -2 & -2 & | & 3 \\ 0 & -3 & -3 & -3 & | & 0 \\ 0 & -2 & 1 & -3 & | & -11 \end{bmatrix}$ $\begin{array}{l} -R2 + R1 \to R1 \\ \\ 3R2 + R3 \to R3 \\ 2R2 + R4 \to R4 \end{array}$

$\begin{bmatrix} 1 & 0 & 3 & 3 & | & 1 \\ 0 & 1 & -2 & -2 & | & 3 \\ 0 & 0 & -9 & -9 & | & 9 \\ 0 & 0 & -3 & -7 & | & -5 \end{bmatrix}$ $(-1/9)R3 \to R3$

$\begin{bmatrix} 1 & 0 & 3 & 3 & | & 1 \\ 0 & 1 & -2 & -2 & | & 3 \\ 0 & 0 & 1 & 1 & | & -1 \\ 0 & 0 & -3 & -7 & | & -5 \end{bmatrix}$ $\begin{array}{l} -3R3 + R1 \to R1 \\ 2R3 + R2 \to R2 \\ \\ 3R3 + R4 \to R4 \end{array}$

$\begin{bmatrix} 1 & 0 & 0 & 0 & | & 4 \\ 0 & 1 & 0 & 0 & | & 1 \\ 0 & 0 & 1 & 1 & | & -1 \\ 0 & 0 & 0 & -4 & | & -8 \end{bmatrix}$ $(-1/4)R4 \to R4$

$\begin{bmatrix} 1 & 0 & 0 & 0 & | & 4 \\ 0 & 1 & 0 & 0 & | & 1 \\ 0 & 0 & 1 & 1 & | & -1 \\ 0 & 0 & 0 & 1 & | & 2 \end{bmatrix}$ $-R4 + R3 \to R3$

23

$$\begin{bmatrix} 1 & 0 & 0 & 0 & | & 4 \\ 0 & 1 & 0 & 0 & | & 1 \\ 0 & 0 & 1 & 0 & | & -3 \\ 0 & 0 & 0 & 1 & | & 2 \end{bmatrix} \qquad (4, \ 1, \ -3, \ 2)$$

51.
$$\begin{bmatrix} 2 & 6 & 4 & -2 & | & 18 \\ 1 & 4 & -2 & -1 & | & -1 \\ 3 & -1 & -1 & 2 & | & 6 \\ -1 & -2 & -5 & 0 & | & -20 \end{bmatrix} \qquad \begin{bmatrix} 1 & 3 & 2 & -1 & | & 9 \\ 0 & 1 & -4 & 0 & | & -10 \\ 0 & -10 & -7 & 5 & | & -21 \\ 0 & 1 & -3 & -1 & | & -11 \end{bmatrix}$$

$$\begin{bmatrix} 1 & 0 & 14 & -1 & | & 39 \\ 0 & 1 & -4 & 0 & | & -10 \\ 0 & 0 & -47 & 5 & | & -121 \\ 0 & 0 & 1 & -1 & | & -1 \end{bmatrix} \qquad \begin{bmatrix} 1 & 0 & 14 & -1 & | & 39 \\ 0 & 1 & -4 & 0 & | & -10 \\ 0 & 0 & 1 & -1 & | & -1 \\ 0 & 0 & -47 & 5 & | & -121 \end{bmatrix}$$

$$\begin{bmatrix} 1 & 0 & 0 & 13 & | & 53 \\ 0 & 1 & 0 & -4 & | & -14 \\ 0 & 0 & 1 & -1 & | & -1 \\ 0 & 0 & 0 & -42 & | & -168 \end{bmatrix} \qquad \begin{bmatrix} 1 & 0 & 0 & 0 & | & 1 \\ 0 & 1 & 0 & 0 & | & 2 \\ 0 & 0 & 1 & 0 & | & 3 \\ 0 & 0 & 0 & 1 & | & 4 \end{bmatrix} \qquad (1, \ 2, \ 3, \ 4)$$

53. Let x_1 = number cases of Regular, x_2 = number cases of Premium, x_3 = number cases of Classic.
$$4x_1 + 4x_2 + 5x_3 = 316 \quad \text{(Apple juice)}$$
$$5x_1 + 4x_2 + 2x_3 = 292 \quad \text{(Pineapple juice)}$$
$$x_1 + 2x_2 + 3x_3 = 142 \quad \text{(Cranberry juice)}$$

55. Let x_1 = number of student tickets, x_2 = number of faculty tickets, x_3 = number of general public tickets.
$$3x_1 + 5x_2 + 8x_3 = 2542$$
$$x_1 = 3x_2$$
$$x_3 = 2x_1$$

57. Let x = number shares of X, y = number shares of Y, z = number shares of Z.
$$44x + 22y + 64z = 20{,}480$$
$$42x + 28y + 62z = 20{,}720$$
$$42x + 30y + 60z = 20{,}580$$

59. Let x_1 = number of Portfolio I, x_2 = number of Portfolio II, x_3 = number of Portfolio III.
$$x_1 + 3x_2 + 2x_3 = 15 \quad \text{(Blue chip)}$$
$$3x_1 + 2x_2 + x_3 = 14 \quad \text{(Bonds)}$$
$$x_1 + 2x_2 + 4x_3 = 16 \quad \text{(High risk)}$$

$$\begin{bmatrix} 1 & 3 & 2 & | & 15 \\ 3 & 2 & 1 & | & 14 \\ 1 & 2 & 4 & | & 16 \end{bmatrix} \qquad \begin{bmatrix} 1 & 3 & 2 & | & 15 \\ 0 & -7 & -5 & | & -31 \\ 0 & -1 & 2 & | & 1 \end{bmatrix} \qquad \begin{bmatrix} 1 & 0 & 8 & | & 18 \\ 0 & 1 & -2 & | & -1 \\ 0 & 0 & -19 & | & -38 \end{bmatrix}$$

$$\begin{bmatrix} 1 & 0 & 8 & | & 18 \\ 0 & 1 & -2 & | & -1 \\ 0 & 0 & 1 & | & 2 \end{bmatrix} \qquad \begin{bmatrix} 1 & 0 & 0 & | & 2 \\ 0 & 1 & 0 & | & 3 \\ 0 & 0 & 1 & | & 2 \end{bmatrix}$$

2 of Portfolio I, 3 of Portfolio II, 2 of Portfolio III

61. Let A = amount invested in stock A, B = amount invested in stock B, C = amount invested in stock C.

$$A + B + C = 40,000$$
$$.06A + .07B + .08C = 2730 \quad \text{(Dividends)}$$
$$.03A + .04B + .02C = 1080 \quad \text{(Increase)}$$

$$\begin{bmatrix} 1 & 1 & 1 & | & 40000 \\ .06 & .07 & .08 & | & 2730 \\ .03 & .04 & .02 & | & 1080 \end{bmatrix} \begin{bmatrix} 1 & 1 & 1 & | & 40000 \\ 0 & 1 & 2 & | & 33000 \\ 0 & 1 & -1 & | & -12000 \end{bmatrix} \begin{bmatrix} 1 & 0 & -1 & | & 7000 \\ 0 & 1 & 2 & | & 33000 \\ 0 & 0 & -3 & | & -45000 \end{bmatrix}$$

$$\begin{bmatrix} 1 & 0 & -1 & | & 7000 \\ 0 & 1 & 2 & | & 33000 \\ 0 & 0 & 1 & | & 15000 \end{bmatrix} \begin{bmatrix} 1 & 0 & 0 & | & 22000 \\ 0 & 1 & 0 & | & 3000 \\ 0 & 0 & 1 & | & 15000 \end{bmatrix}$$

$22,000 in stock A, $3,000 in stock B, $15,000 in stock C

63. Let x_1 = number high school tickets, x_2 = number college tickets, x_3 = number adult tickets.

$$4x_1 + 6x_2 + 8x_3 = 14980 \quad \text{(Rock Stars)}$$
$$4x_1 + 5x_2 + 9x_3 = 14430 \quad \text{(Smooth Sounds)}$$
$$2x_1 + 7x_2 + 7x_3 = 14450 \quad \text{(Baroque Band)}$$

$$\begin{bmatrix} 4 & 6 & 8 & | & 14980 \\ 4 & 5 & 9 & | & 14430 \\ 2 & 7 & 7 & | & 14450 \end{bmatrix} \begin{bmatrix} 1 & 3/2 & 2 & | & 3745 \\ 0 & -1 & 1 & | & -550 \\ 0 & 4 & 3 & | & 6960 \end{bmatrix} \begin{bmatrix} 1 & 0 & 7/2 & | & 2920 \\ 0 & 1 & -1 & | & 550 \\ 0 & 0 & 7 & | & 4760 \end{bmatrix}$$

$$\begin{bmatrix} 1 & 0 & 7/2 & | & 2920 \\ 0 & 1 & -1 & | & 550 \\ 0 & 0 & 1 & | & 680 \end{bmatrix} \begin{bmatrix} 1 & 0 & 0 & | & 540 \\ 0 & 1 & 0 & | & 1230 \\ 0 & 0 & 1 & | & 680 \end{bmatrix}$$

540 to high school students, 1230 to college students, 680 to adults

65.
$$\begin{bmatrix} 1 & 2 & 3 & | & 142 \\ 5 & 4 & 2 & | & 292 \\ 4 & 4 & 5 & | & 316 \end{bmatrix} \begin{bmatrix} 1 & 2 & 3 & | & 142 \\ 0 & -6 & -13 & | & -418 \\ 0 & -4 & -7 & | & -252 \end{bmatrix} \begin{bmatrix} 1 & 0 & -4/3 & | & 8/3 \\ 0 & 1 & 13/6 & | & 209/3 \\ 0 & 0 & 5/3 & | & 80/3 \end{bmatrix}$$

$$\begin{bmatrix} 1 & 0 & -4/3 & | & 8/3 \\ 0 & 1 & 13/6 & | & 209/3 \\ 0 & 0 & 1 & | & 16 \end{bmatrix} \begin{bmatrix} 1 & 0 & 0 & | & 24 \\ 0 & 1 & 0 & | & 35 \\ 0 & 0 & 1 & | & 16 \end{bmatrix} \begin{matrix} 24 \text{ cases of Regular} \\ 35 \text{ cases of Premium} \\ 16 \text{ cases of Classic} \end{matrix}$$

67.
$$\begin{bmatrix} 1 & -3 & 0 & | & 0 \\ -2 & 0 & 1 & | & 0 \\ 3 & 5 & 8 & | & 2542 \end{bmatrix} \begin{bmatrix} 1 & -3 & 0 & | & 0 \\ 0 & -6 & 1 & | & 0 \\ 0 & 14 & 8 & | & 2542 \end{bmatrix}$$

$$\begin{bmatrix} 1 & 0 & -1/2 & | & 0 \\ 0 & 1 & -1/6 & | & 0 \\ 0 & 0 & 31/3 & | & 2542 \end{bmatrix} \begin{bmatrix} 1 & 0 & 0 & | & 123 \\ 0 & 1 & 0 & | & 41 \\ 0 & 0 & 1 & | & 246 \end{bmatrix}$$

123 students, 41 faculty, 246 general public

69.
$$\begin{bmatrix} 44 & 22 & 64 & | & 20480 \\ 42 & 28 & 62 & | & 20720 \\ 42 & 30 & 60 & | & 20580 \end{bmatrix} \begin{bmatrix} 1 & 1/2 & 16/11 & | & 5120/11 \\ 0 & 7 & 10/11 & | & 12880/11 \\ 0 & 9 & -12/11 & | & 11340/11 \end{bmatrix}$$

$$\begin{bmatrix} 1 & 0 & 107/77 & | & 4200/11 \\ 0 & 1 & 10/77 & | & 1840/11 \\ 0 & 0 & -174/77 & | & -5220/11 \end{bmatrix} \begin{bmatrix} 1 & 1 & 107/77 & | & 4200/11 \\ 0 & 1 & 10/77 & | & 1840/11 \\ 0 & 0 & 1 & | & 210 \end{bmatrix}$$

$$\begin{bmatrix} 1 & 0 & 0 & | & 90 \\ 0 & 1 & 0 & | & 140 \\ 0 & 0 & 1 & | & 210 \end{bmatrix} \quad \begin{matrix} 90 \text{ shares of X} \\ 140 \text{ shares of Y} \\ 210 \text{ shares of Z} \end{matrix}$$

71.   Let x_1 = number of days A operates, x_2 = number of days B
      operates, x_3 = number of days C operates.
$$300x_1 + 700x_2 + 400x_3 = 39500 \text{ (Interior)}$$
$$500x_1 + 900x_2 + 400x_3 = 52500 \text{ (Exterior)}$$
$$200x_1 + 100x_2 + 800x_3 = 12500 \text{ (Rough Cedar)}$$

75.   (a) $\begin{bmatrix} 1.1 & 1.2 & -1.3 & 29.76 \\ 3.5 & 4.1 & -2.2 & 81.94 \\ 2.3 & 0 & 1.4 & 17.70 \end{bmatrix}$ (b) (5.149, -0.990)

Section 2.3

1.   In reduced echelon form

3.   Not in reduced echelon form because column 3 does not contain a
     zero in row 2.

5.   In reduced echelon form.

7.   Not in reduced echelon form because the leading 1 in row 3 is to
     the left of the leading 1 in row 2, and the 3 in row 3 should be 0.

9.   Perform the following row operations to reduce the third column.
     $\frac{1}{4}$ R3 → R3

     -2R3 + R1 → R1
     -3R3 + R2 → R2   and obtain
$$\begin{bmatrix} 1 & 0 & 0 & | & 1 \\ 0 & 1 & 0 & | & -8 \\ 0 & 0 & 1 & | & 2 \end{bmatrix}$$

11. Exchange rows 2 and 3, then get a zero in the (1, 2) position with
     R2 ↔ R3
     -2R2 + R1 → R1
     giving $\begin{bmatrix} 1 & 0 & -1 & | & -8 \\ 0 & 1 & 2 & | & 6 \\ 0 & 0 & 5 & | & -3 \end{bmatrix}$

13. Get a 1 in the (2, 2) position and 0 in (1, 2) position with
     R2 ↔ R3
     -3R2 + R1 → R1
$$\begin{bmatrix} 1 & 0 & -13 & 10 & | & -19 \\ 0 & 1 & 4 & -2 & | & 7 \\ 0 & 0 & 2 & 3 & | & 5 \end{bmatrix}$$

15. $\begin{bmatrix} 1 & 2 & -3 & | & 2 \\ 1 & 0 & 3 & | & -2 \\ 3 & 5 & -7 & | & 2 \end{bmatrix}$ $\begin{bmatrix} 1 & 2 & -3 & | & 2 \\ 0 & -2 & 6 & | & -4 \\ 0 & -1 & 2 & | & -4 \end{bmatrix}$

     $\begin{bmatrix} 1 & 0 & 3 & | & -2 \\ 0 & 1 & -3 & | & 2 \\ 0 & 0 & -1 & | & -2 \end{bmatrix}$ $\begin{bmatrix} 1 & 0 & 0 & | & -8 \\ 0 & 1 & 0 & | & 8 \\ 0 & 0 & 1 & | & 2 \end{bmatrix}$

17.
$$\begin{bmatrix} 1 & 3 & 2 & \bigm| & 1 \\ 3 & -1 & 4 & \bigm| & 9 \\ 2 & -4 & 2 & \bigm| & 8 \end{bmatrix} \quad \begin{bmatrix} 1 & 3 & 2 & \bigm| & 1 \\ 0 & -10 & -2 & \bigm| & 6 \\ 0 & -10 & -2 & \bigm| & 6 \end{bmatrix} \quad \begin{bmatrix} 1 & 0 & 7/5 & \bigm| & 14/5 \\ 0 & 1 & 1/5 & \bigm| & -3/5 \\ 0 & 0 & 0 & \bigm| & 0 \end{bmatrix}$$

19.
$$\begin{bmatrix} 1 & -1 & 3 & 1 & \bigm| & 0 \\ 2 & -2 & 7 & 0 & \bigm| & -5 \\ 1 & -1 & 2 & 1 & \bigm| & -1 \\ -2 & 2 & 6 & 2 & \bigm| & -1 \end{bmatrix} \quad \begin{bmatrix} 1 & -1 & 3 & 1 & \bigm| & 0 \\ 0 & 0 & 1 & -2 & \bigm| & -5 \\ 0 & 0 & -1 & 0 & \bigm| & -1 \\ 0 & 0 & 12 & 4 & \bigm| & -1 \end{bmatrix}$$

$$\begin{bmatrix} 1 & -1 & 0 & 7 & \bigm| & 15 \\ 0 & 0 & 1 & -2 & \bigm| & -5 \\ 0 & 0 & 0 & -2 & \bigm| & -6 \\ 0 & 0 & 0 & 28 & \bigm| & 59 \end{bmatrix} \quad \begin{bmatrix} 1 & -1 & 0 & 0 & \bigm| & -6 \\ 0 & 0 & 1 & 0 & \bigm| & 1 \\ 0 & 0 & 0 & 1 & \bigm| & 3 \\ 0 & 0 & 0 & 0 & \bigm| & -25 \end{bmatrix}$$

$$\begin{bmatrix} 1 & -1 & 0 & 0 & \bigm| & 0 \\ 0 & 0 & 1 & 0 & \bigm| & 0 \\ 0 & 0 & 0 & 1 & \bigm| & 0 \\ 0 & 0 & 0 & 0 & \bigm| & 1 \end{bmatrix}$$

21. $x_1 = 3$
      $x_2 = -2$
      $x_3 = 5$ The solution is $(3, -2, 5)$.

23. $x_1 \quad + 3x_3 = 4$
      $\quad x_2 + \quad x_3 = -6$
      $\qquad\qquad x_4 = 2$
      Infinite number of solutions of the form $(4 - 3k, -6 - k, k, 2)$.

25. $x_1 = 0$
      $x_2 = 0$
      $0 = 1$
      No solution

27. $x_1 \quad + 3x_4 = 0$
      $\quad x_2 \quad - 2x_4 = 0$
      $\qquad x_3 + 7x_4 = 0$
      $\qquad\qquad\qquad 0 = 0$
      Infinite number of solutions
      of the form
      $(-3k, 2k, -7k, k)$.

29.
$$\begin{bmatrix} 1 & 4 & -2 & \bigm| & 13 \\ 3 & -1 & 4 & \bigm| & 6 \\ 2 & -5 & 6 & \bigm| & -4 \end{bmatrix} \quad \begin{bmatrix} 1 & 4 & -2 & \bigm| & 13 \\ 0 & -13 & 10 & \bigm| & -33 \\ 0 & -13 & 10 & \bigm| & -30 \end{bmatrix}$$

$$\begin{bmatrix} 1 & 0 & 14/13 & \bigm| & 37/13 \\ 0 & 1 & -10/13 & \bigm| & 33/13 \\ 0 & 0 & 0 & \bigm| & 3 \end{bmatrix} \qquad \text{No solution}$$

31.
$$\begin{bmatrix} 3 & -2 & 2 & | & 10 \\ 2 & 1 & 3 & | & 3 \\ 1 & 1 & -1 & | & 5 \end{bmatrix} \qquad \begin{bmatrix} 1 & 1 & -1 & | & 5 \\ 2 & 1 & 3 & | & 3 \\ 3 & -2 & 2 & | & 10 \end{bmatrix}$$

$$\begin{bmatrix} 1 & 1 & -1 & | & 5 \\ 0 & -1 & 5 & | & -7 \\ 0 & -5 & 5 & | & -5 \end{bmatrix} \qquad \begin{bmatrix} 1 & 0 & 4 & | & -2 \\ 0 & 1 & -5 & | & 7 \\ 0 & 0 & -20 & | & 30 \end{bmatrix}$$

$$\begin{bmatrix} 1 & 0 & 0 & | & 4 \\ 0 & 1 & 0 & | & -1/2 \\ 0 & 0 & 1 & | & -3/2 \end{bmatrix} \qquad x_1 = 4, \quad x_2 = -1/2, \quad x_3 = -3/2$$

33.
$$\begin{bmatrix} 1 & 1 & 1 & -1 & | & -3 \\ 2 & 3 & 1 & -5 & | & -9 \\ 1 & 3 & -1 & -6 & | & 7 \end{bmatrix} \qquad \begin{bmatrix} 1 & 1 & 1 & -1 & | & -3 \\ 0 & 1 & -1 & -3 & | & -3 \\ 0 & 2 & -2 & -5 & | & 10 \end{bmatrix}$$

$$\begin{bmatrix} 1 & 0 & 2 & 2 & | & 0 \\ 0 & 1 & -1 & -3 & | & -3 \\ 0 & 0 & 0 & 1 & | & 16 \end{bmatrix} \qquad \begin{bmatrix} 1 & 0 & 2 & 0 & | & -32 \\ 0 & 1 & -1 & 0 & | & 45 \\ 0 & 0 & 0 & 1 & | & 16 \end{bmatrix}$$

$$x_1 = -32 - 2x_3, \quad x_2 = 45 + x_3, \quad x_4 = 16$$
or $(-32 - 2k, 45 + k, k, 16)$

35.
$$\begin{bmatrix} 1 & -1 & 1 & | & 3 \\ -2 & 3 & 1 & | & -8 \\ 4 & -2 & 10 & | & 10 \end{bmatrix} \qquad \begin{bmatrix} 1 & -1 & 1 & | & 3 \\ 0 & 1 & 3 & | & -2 \\ 0 & 2 & 6 & | & -2 \end{bmatrix}$$

$$\begin{bmatrix} 1 & 0 & 4 & | & 1 \\ 0 & 1 & 3 & | & -2 \\ 0 & 0 & 0 & | & 2 \end{bmatrix} \qquad \text{No solution}$$

37.
$$\begin{bmatrix} 1 & 1 & 1 & | & 6 \\ 1 & -3 & 2 & | & 1 \\ 3 & -1 & 4 & | & 5 \end{bmatrix} \qquad \begin{bmatrix} 1 & 1 & 1 & | & 6 \\ 0 & -4 & 1 & | & -5 \\ 0 & -4 & 1 & | & -13 \end{bmatrix}$$

$$\begin{bmatrix} 1 & 1 & 1 & | & 6 \\ 0 & -4 & 1 & | & -5 \\ 0 & 0 & 0 & | & -8 \end{bmatrix} \qquad \text{No solution}$$

39.
$$\begin{bmatrix} 1 & 2 & -1 & | & -13 \\ 2 & 5 & 3 & | & -3 \end{bmatrix} \qquad \begin{bmatrix} 1 & 2 & -1 & | & -13 \\ 0 & 1 & 5 & | & 23 \end{bmatrix}$$

$$\begin{bmatrix} 1 & 0 & -11 & | & -59 \\ 0 & 1 & 5 & | & 23 \end{bmatrix} \qquad x_1 = -59 + 11x_3, \quad x_2 = 23 - 5x_3$$
or $(-59 + 11k, 23 - 5k, k)$

41.
$$\begin{bmatrix} 1 & -2 & 1 & 1 & -2 & | & -9 \\ 5 & 1 & -6 & -6 & 1 & | & 21 \end{bmatrix} \qquad \begin{bmatrix} 1 & -2 & 1 & 1 & -2 & | & -9 \\ 0 & 11 & -11 & -11 & 11 & | & 66 \end{bmatrix}$$

$$\begin{bmatrix} 1 & 0 & -1 & -1 & 0 & | & 3 \\ 0 & 1 & -1 & -1 & 1 & | & 6 \end{bmatrix} \qquad x_1 = 3 + x_3 + x_4, \quad x_2 = 6 + x_3 + x_4 - x_5$$
or $(3 + k + r, 6 + k + r - s, k, r, s)$

43.
$$\begin{bmatrix} 1 & 1 & -3 & 1 & | & 4 \\ -2 & -2 & 6 & -2 & | & 3 \end{bmatrix} \qquad \begin{bmatrix} 1 & 1 & -3 & 1 & | & 4 \\ 0 & 0 & 0 & 0 & | & 11 \end{bmatrix}$$

No solution

45.
$$\left[\begin{array}{rrrr|r} 1 & 1 & -1 & -1 & -1 \\ 3 & -2 & -4 & 2 & 1 \\ 4 & -1 & -5 & 1 & 5 \end{array}\right] \qquad \left[\begin{array}{rrrr|r} 1 & 1 & -1 & -1 & -1 \\ 0 & -5 & -1 & 5 & 4 \\ 0 & -5 & -1 & 5 & 9 \end{array}\right]$$

$$\left[\begin{array}{rrrr|r} 1 & 1 & -1 & -1 & -1 \\ 0 & -5 & -1 & 5 & 4 \\ 0 & 0 & 0 & 0 & 5 \end{array}\right] \qquad \text{No solution}$$

47.
$$\left[\begin{array}{rr|r} 1 & 4 & -10 \\ -2 & 3 & -13 \\ 5 & -2 & 16 \end{array}\right] \quad \left[\begin{array}{rr|r} 1 & 4 & -10 \\ 0 & 11 & -33 \\ 0 & -22 & 66 \end{array}\right] \quad \left[\begin{array}{rr|r} 1 & 0 & 2 \\ 0 & 1 & -3 \\ 0 & 0 & 0 \end{array}\right] \qquad (2, -3)$$

49.
$$\left[\begin{array}{rr|r} 1 & -1 & -7 \\ 1 & 1 & -3 \\ 3 & -1 & -17 \end{array}\right] \quad \left[\begin{array}{rr|r} 1 & -1 & -7 \\ 0 & 2 & 4 \\ 0 & 2 & 4 \end{array}\right] \quad \left[\begin{array}{rr|r} 1 & 0 & -5 \\ 0 & 1 & 2 \\ 0 & 0 & 0 \end{array}\right]$$

$x = -5, \ y = 2$

51.
$$\left[\begin{array}{rrr|r} 0 & 1 & 2 & 7 \\ 1 & -2 & -6 & -18 \\ 1 & -1 & -2 & -5 \\ 2 & -5 & -15 & -46 \end{array}\right] \quad \left[\begin{array}{rrr|r} 1 & -2 & -6 & -18 \\ 0 & 1 & 2 & 7 \\ 1 & -1 & -2 & -5 \\ 2 & -5 & -15 & -46 \end{array}\right] \quad \left[\begin{array}{rrr|r} 1 & -2 & -6 & -18 \\ 0 & 1 & 2 & 7 \\ 0 & 1 & 4 & 13 \\ 0 & -1 & -3 & -10 \end{array}\right]$$

$$\left[\begin{array}{rrr|r} 1 & 0 & -2 & -4 \\ 0 & 1 & 2 & 7 \\ 0 & 0 & 2 & 6 \\ 0 & 0 & -1 & -3 \end{array}\right] \quad \left[\begin{array}{rrr|r} 1 & 0 & 0 & 2 \\ 0 & 1 & 0 & 1 \\ 0 & 0 & 1 & 3 \\ 0 & 0 & 0 & 0 \end{array}\right] \qquad (2, 1, 3)$$

53.
$$\left[\begin{array}{rrr|r} 3 & -2 & 4 & 4 \\ 2 & 5 & -1 & -2 \\ 1 & -7 & 5 & 6 \\ 5 & 3 & 3 & 3 \end{array}\right] \qquad \left[\begin{array}{rrr|r} 1 & -7 & 5 & 6 \\ 3 & -2 & 4 & 4 \\ 2 & 5 & -1 & -2 \\ 5 & 3 & 3 & 3 \end{array}\right]$$

$$\left[\begin{array}{rrr|r} 1 & -7 & 5 & 6 \\ 0 & 19 & -11 & -14 \\ 0 & 19 & -11 & -14 \\ 0 & 38 & -22 & -27 \end{array}\right] \quad \left[\begin{array}{rrr|r} 1 & 0 & 18/19 & 16/19 \\ 0 & 1 & -11/19 & -14/19 \\ 0 & 0 & 0 & 0 \\ 0 & 0 & 0 & 1 \end{array}\right]$$

55.
$$\left[\begin{array}{rr|r} 2 & -5 & 5 \\ 6 & 1 & 31 \\ 2 & 11 & 18 \end{array}\right] \quad \left[\begin{array}{rr|r} 1 & -5/2 & 5/2 \\ 0 & 16 & 16 \\ 0 & 16 & 13 \end{array}\right] \quad \left[\begin{array}{rr|r} 1 & -5/2 & 5/2 \\ 0 & 16 & 16 \\ 0 & 0 & -3 \end{array}\right]$$

No solution

57.
$$\left[\begin{array}{rrrr|r} 1 & -1 & 2 & 0 & 7 \\ 3 & -4 & 18 & -13 & 17 \\ 2 & -2 & 2 & -4 & 12 \\ -1 & 1 & -1 & 2 & -6 \\ -3 & 1 & -8 & -10 & -21 \end{array}\right] \quad \left[\begin{array}{rrrr|r} 1 & -1 & 2 & 0 & 7 \\ 0 & -1 & 12 & -13 & -4 \\ 0 & 0 & -2 & -4 & -2 \\ 0 & 0 & 1 & 2 & 1 \\ 0 & -2 & -2 & -10 & 0 \end{array}\right]$$

$$\left[\begin{array}{rrrr|r} 1 & 0 & -10 & 13 & 11 \\ 0 & 1 & -12 & 13 & 4 \\ 0 & 0 & -2 & -4 & -2 \\ 0 & 0 & 1 & 2 & 1 \\ 0 & 0 & -26 & 16 & 8 \end{array}\right] \quad \left[\begin{array}{rrrr|r} 1 & 0 & 0 & 33 & 21 \\ 0 & 1 & 0 & 37 & 16 \\ 0 & 0 & 1 & 2 & 1 \\ 0 & 0 & 0 & 0 & 0 \\ 0 & 0 & 0 & 68 & 34 \end{array}\right]$$

$$\begin{bmatrix} 1 & 0 & 0 & 0 & \bigm| & 9/2 \\ 0 & 1 & 0 & 0 & \bigm| & -5/2 \\ 0 & 0 & 1 & 0 & \bigm| & 0 \\ 0 & 0 & 0 & 0 & \bigm| & 0 \\ 0 & 0 & 0 & 1 & \bigm| & 1/2 \end{bmatrix} \qquad x_1 = \frac{9}{2},\ x_2 = -\frac{5}{2},\ x_3 = 0,\ x_4 = \frac{1}{2}$$

59. $\begin{bmatrix} 1 & 2 & -1 & -1 & \bigm| & 0 \\ 1 & 2 & 0 & 1 & \bigm| & 4 \\ -1 & -2 & 2 & 4 & \bigm| & 5 \\ -1 & -1 & -1 & 0 & \bigm| & 1 \end{bmatrix} \quad \begin{bmatrix} 1 & 2 & -1 & -1 & \bigm| & 0 \\ 0 & 0 & 1 & 2 & \bigm| & 4 \\ 0 & 0 & 1 & 3 & \bigm| & 5 \\ 0 & 1 & -2 & -1 & \bigm| & 1 \end{bmatrix} \quad \begin{bmatrix} 1 & 0 & 3 & 1 & \bigm| & -2 \\ 0 & 1 & -2 & -1 & \bigm| & 1 \\ 0 & 0 & 1 & 3 & \bigm| & 5 \\ 0 & 0 & 1 & 2 & \bigm| & 4 \end{bmatrix}$

$\begin{bmatrix} 1 & 0 & 0 & -8 & \bigm| & -17 \\ 0 & 1 & 0 & 5 & \bigm| & 11 \\ 0 & 0 & 1 & 3 & \bigm| & 5 \\ 0 & 0 & 0 & -1 & \bigm| & -1 \end{bmatrix} \quad \begin{bmatrix} 1 & 0 & 0 & 0 & \bigm| & -9 \\ 0 & 1 & 0 & 0 & \bigm| & 6 \\ 0 & 0 & 1 & 0 & \bigm| & 2 \\ 0 & 0 & 0 & 1 & \bigm| & 1 \end{bmatrix} \qquad (-9,\ 6,\ 2,\ 1)$

61. $\begin{bmatrix} 1 & 2 & -3 & 2 & 5 & -1 & \bigm| & 0 \\ -2 & -4 & 6 & -1 & -4 & 5 & \bigm| & 0 \\ 3 & 6 & -9 & 5 & 13 & -4 & \bigm| & 0 \end{bmatrix} \qquad \begin{bmatrix} 1 & 2 & -3 & 2 & 5 & -1 & \bigm| & 0 \\ 0 & 0 & 0 & 3 & 6 & 3 & \bigm| & 0 \\ 0 & 0 & 0 & -1 & -2 & -1 & \bigm| & 0 \end{bmatrix}$

$\begin{bmatrix} 1 & 2 & -3 & 0 & 1 & -3 & \bigm| & 0 \\ 0 & 0 & 0 & 1 & 2 & 1 & \bigm| & 0 \\ 0 & 0 & 0 & 0 & 0 & 0 & \bigm| & 0 \end{bmatrix}$

$x_1 = -2x_2 + 3x_3 - x_5 + 3x_6,\quad x_4 = -2x_5 - x_6$

or $(-2k + 3r - s + 3t,\ k,\ r,\ -2s - t,\ s,\ t)$

63. Let x_1 = amount in stocks, x_2 = amount in bonds,
x_3 = amount in money markets.

$x_1 + x_2 + x_3 = 45{,}000$ (Total)

$-2x_1 + x_2 + x_3 = 0$

$.10x_1 + .07x_2 + .075x_3 = 3{,}660$ (Return)

$\begin{bmatrix} 1 & 1 & 1 & \bigm| & 45{,}000 \\ -2 & 1 & 1 & \bigm| & 0 \\ .10 & .07 & .075 & \bigm| & 3{,}660 \end{bmatrix} \qquad \begin{bmatrix} 1 & 1 & 1 & \bigm| & 45{,}000 \\ 0 & 3 & 3 & \bigm| & 90{,}000 \\ 0 & -30 & -25 & \bigm| & -840{,}000 \end{bmatrix}$

$\begin{bmatrix} 1 & 0 & 0 & \bigm| & 15{,}000 \\ 0 & 1 & 1 & \bigm| & 30{,}000 \\ 0 & 0 & 5 & \bigm| & 60{,}000 \end{bmatrix} \qquad \begin{bmatrix} 1 & 0 & 0 & \bigm| & 15{,}000 \\ 0 & 1 & 0 & \bigm| & 18{,}000 \\ 0 & 0 & 1 & \bigm| & 12{,}000 \end{bmatrix}$

$\$15{,}000$ in stocks, $\$18{,}000$ in bonds, $\$12{,}000$ in money market

65. Let x = minutes jogging, y = minutes playing handball,
z = minutes biking.

$x + y + z = 60$

$13x + 11y + 7z = 660$

$x - 2z = 0$

$\begin{bmatrix} 1 & 1 & 1 & \bigm| & 60 \\ 13 & 11 & 7 & \bigm| & 660 \\ 1 & 0 & -2 & \bigm| & 0 \end{bmatrix} \qquad \begin{bmatrix} 1 & 1 & 1 & \bigm| & 60 \\ 0 & -2 & -6 & \bigm| & -120 \\ 0 & -1 & -3 & \bigm| & -60 \end{bmatrix}$

$\begin{bmatrix} 1 & 0 & -2 & \bigm| & 0 \\ 0 & 1 & 3 & \bigm| & 60 \\ 0 & 0 & 0 & \bigm| & 0 \end{bmatrix} \qquad x = 2z,\quad y = 60 - 3z$

She may bike from 0 to 20 minutes, then should jog for twice that
time, and play handball for the rest of the 60 minutes.

67. Let x_1 = hours of Math
       x_2 = hours of English
       x_3 = hours of Chemistry
       x_4 = hours of History

$$x_1 + x_2 + x_3 + x_4 = 42$$
$$x_1 + x_2 = 21$$
$$x_1 = 2x_2$$
$$x_2 = 2x_4$$

$$\left[\begin{array}{cccc|c} 1 & 1 & 1 & 1 & 42 \\ 1 & 1 & 0 & 0 & 21 \\ 1 & -2 & 0 & 0 & 0 \\ 0 & 1 & 0 & -2 & 0 \end{array}\right] \qquad \left[\begin{array}{cccc|c} 1 & 1 & 1 & 1 & 42 \\ 0 & 0 & -1 & -1 & -21 \\ 0 & -3 & -1 & -1 & -42 \\ 0 & 1 & 0 & -2 & 0 \end{array}\right]$$

$$\left[\begin{array}{cccc|c} 1 & 1 & 1 & 1 & 42 \\ 0 & 1 & 1/3 & 1/3 & 14 \\ 0 & 0 & 1 & 1 & 21 \\ 0 & 1 & 0 & -2 & 0 \end{array}\right] \qquad \left[\begin{array}{cccc|c} 1 & 0 & 2/3 & 2/3 & 28 \\ 0 & 1 & 1/3 & 1/3 & 14 \\ 0 & 0 & 1 & 1 & 21 \\ 0 & 0 & -1/3 & -7/3 & -14 \end{array}\right]$$

$$\left[\begin{array}{cccc|c} 1 & 0 & 0 & 0 & 14 \\ 0 & 1 & 0 & 0 & 7 \\ 0 & 0 & 1 & 1 & 21 \\ 0 & 0 & 0 & -2 & -7 \end{array}\right] \qquad \left[\begin{array}{cccc|c} 1 & 0 & 0 & 0 & 14 \\ 0 & 1 & 0 & 0 & 7 \\ 0 & 0 & 1 & 0 & 17.5 \\ 0 & 0 & 0 & 1 & 3.5 \end{array}\right]$$

14 hours for Math, 7 hours for English, and 17.5 hours for Chemistry and 3.5 hours for History.

69. (a)   Let x_1 = number supplied by Sweats-Plus to Spirit Shop 1
          x_2 = number supplied by Sweats-Plus to Spirit Shop 2
          x_3 = number supplied by Sweats-Plus to Spirit Shop 3
          x_4 = number supplied by Imprint-Sweats to Spirit Shop 1
          x_5 = number supplied by Imprint-Sweats to Spirit Shop 2
          x_6 = number supplied by Imprint-Sweats to Spirit Shop 3
          The system of equations is
$$x_1 + x_4 = 15$$
$$x_2 + x_5 = 20$$
$$x_3 + x_6 = 30$$
$$x_1 + x_2 + x_3 = 40$$
$$x_4 + x_5 + x_6 = 25$$
          The solution is
$$x_1 = -10 + x_5 + x_6$$
$$x_2 = 20 - x_5$$
$$x_3 = 30 - x_6$$
$$x_4 = 25 - x_5 - x_6$$

(b) $x_5 = 5$, $x_6 = 5$ yields $x_1 = 0$, $x_2 = 15$, $x_3 = 25$, $x_4 = 15$
       $x_5 = 10$, $x_6 = 5$ yields $x_1 = 5$, $x_2 = 10$, $x_3 = 25$, $x_4 = 10$
       $x_5 = 0$, $x_6 = 15$ yields $x_1 = 5$, $x_2 = 20$, $x_3 = 15$, $x_4 = 10$
       Other solutions are possible.

(c) x_5 represents the number of sweatshirts supplied by Imprint-Sweats to Spirit Shop 2. Since $x_2 = 20 - x_5$, x_5 must be no greater than 20 or else x_2 would be a negative number. Thus, $x_5 \leq 20$.

(d)   If Imprint-Sweats supplies no sweatshirts to Spirit Shop 2 and Spirit Shop 3, then $x_5 = 0$ and $x_6 = 0$ so $x_1 = -10$. Thus, the order cannot be filled.

(e)   Since $x_1 = -10 + x_5 + x_6$, in order for $x_1 \geq 0$, then $x_5 + x_6 \geq 10$. Imprint-Sweats must supply a total of 10 or more to Spirit Shops 2 and 3.

(f)   Since $x_2 = 20 - x_5$, and $0 \leq x_5 \leq 20$, then $x_2 \leq 20$. Thus, Sweats-Plus supplies 20 or less to Spirit Shop 2.

71.   The reduced echelon form of the augmented matrix is

$$\begin{bmatrix} 1 & 0 & 7/5 & | & 0 \\ 0 & 1 & -1/5 & | & 0 \\ 0 & 0 & 0 & | & 1 \end{bmatrix}$$

which indicates no solution. In order to have a solution, the 2 in the third equation must be changed so that the bottom row of the reduced matrix is all zeros. Substitute c for 2 in the third equation and reduce the augmented matrix.

The augmented matrix is now

$$\begin{bmatrix} 1 & 2 & 1 & | & 3 \\ 2 & -1 & 3 & | & 2 \\ 3 & 1 & 4 & | & c \end{bmatrix} \quad \text{which reduces to} \quad \begin{bmatrix} 1 & 2 & 1 & | & 3 \\ 0 & -5 & 1 & | & -4 \\ 0 & -5 & 1 & | & c-9 \end{bmatrix}$$

We need only to reduce the matrix further by getting a zero in row 3, column 2.

$$\begin{bmatrix} 1 & 2 & 1 & | & 3 \\ 0 & -5 & 1 & | & -4 \\ 0 & 0 & 0 & | & c-5 \end{bmatrix}$$

Since $c - 5$ must be zero, $c = 5$.
The last equation should be $3x + y + 4z = 5$.

77.   The system has infinitely many solutions.

79.

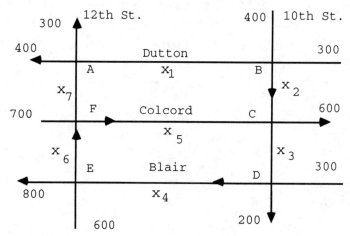

(a)   Label the traffic flow on each block by x_1, x_2, ..., x_7 as shown. Then the condition that incoming traffic equals outgoing traffic at each intersection gives:

At A: $x_1 + x_7 = 700$
At B: $x_1 + x_2 = 700$
At C: $x_2 + x_5 = 600 + x_3$
At D: $x_3 + 300 = 200 + x_4$

$$\text{At E:} \qquad 600 + x_4 = 800 + x_6$$
$$\text{At F:} \qquad x_6 + 700 = x_5 + x_7$$

(b)   The solution to the above system is

$$x_1 = 700 - x_7$$
$$x_2 = x_7$$
$$x_3 = 100 + x_6$$
$$x_4 = 200 + x_6$$
$$x_5 = 700 + x_6 - x_7$$

(c) x_3 represents the traffic flow on 10th St. between Colcord and Blair. Since $x_3 = 100 + x_6$, $x_3 \geq 100$. The minimum traffic flow is 100 vehicles per hour when $x_6 = 0$. The other traffic flows then become

$$x_1 = 700 - x_7$$
$$x_2 = x_7$$
$$x_4 = 200$$
$$x_5 = 700 - x_7$$

81. $x_1 = .2 + 1.6x_3$
$x_2 = -.2 + 1.4x_3$

83. $(-19, 28, -10)$

85. (a) $x_1 = .5 + 3.5x_4$
$x_2 = .5 - 1.83x_4$
$x_3 = 2.33x_4$

(b) $x_2 = .5 - 1.83x_4$ implies $.5 - 1.83x_4 > 0$

so $x_4 < \dfrac{.5}{1.83} = .273$.

Thus, $0 < x_4 < .273$

$x_3 = 2.33x_4$ implies $0 < x_3 < .636$

$x_1 = .5 + 3.5x_4$ implies $x_1 > .5$

and the maximum value of x_1 occurs when x_4 is maximum. Thus, $x_1 < .5 + 3.54(.273) = 1.466$.   so $0.5 < x_1 < 1.466$

$x_2 = .5 - 1.83x_4$ implies $x_2 < .5$ and x_2 is minimum, 0, when x_4 reaches .273 so $0 < x_2 < .5$.

Section 2.4

1.

	Alpha	Beta
Salv. Army	50	65
Boys' Club	85	32
Girl Scouts	68	94

3.

	Joe	Jane	Judy
Checking	12	11	5
Savings	15	18	8
Boxes	8	9	21

5.   2 by 2

7.   3 by 3

9.   4 by 1

11. 2 by 4

13. 2 by 3

15. 2 by 2

17. Not equal

19. Equal

21. Not equal

23. $\begin{bmatrix} 3 & 3 & 2 \\ 7 & 4 & 3 \end{bmatrix}$ 25. $\begin{bmatrix} 2 \\ 58 \end{bmatrix}$ 27. Can <u>not</u> add them

29. $\begin{bmatrix} 6 & 5 & 8 \\ 5 & 1 & 3 \\ 5 & -3 & 4 \end{bmatrix}$ 31. $\begin{bmatrix} 12 & 3 \\ 6 & 15 \end{bmatrix}$ 33. $\begin{bmatrix} 20 \\ 15 \\ 5 \\ 10 \end{bmatrix}$

35. $\begin{bmatrix} -12 & 6 & -15 \end{bmatrix}$ 37. $\begin{bmatrix} 0 & 0 \\ 0 & 0 \end{bmatrix}$

39. (a) $3A = \begin{bmatrix} 3 & 12 \\ -6 & 9 \end{bmatrix}$, $-2B = \begin{bmatrix} 0 & -4 \\ -8 & -2 \end{bmatrix}$, $5C = \begin{bmatrix} 5 & -10 \\ 5 & -15 \end{bmatrix}$

   (b) $A + C = \begin{bmatrix} 2 & 2 \\ -1 & 0 \end{bmatrix}$ (c) $3A - 2B = \begin{bmatrix} 3 & 8 \\ -14 & 7 \end{bmatrix}$

   (d) $A - 2B + 5C = \begin{bmatrix} 1 & 4 \\ -2 & 3 \end{bmatrix} + \begin{bmatrix} 0 & -4 \\ -8 & -2 \end{bmatrix} + \begin{bmatrix} 5 & -10 \\ 5 & -15 \end{bmatrix}$

   $= \begin{bmatrix} 6 & -10 \\ -5 & -14 \end{bmatrix}$

41.
	I	II	III
PC	23	17	20
Print	19	22	11
Disk	151	151	105

43. $x = 3$

45. $6x + 4 = 14x - 13$ so $x = \dfrac{17}{8}$

47.
$\dfrac{1}{12} A = \begin{bmatrix} 65/12 & 55/6 & 20/3 \\ 30/4 & 45/4 & 5 \\ 25/4 & 28/3 & 7 \end{bmatrix}$ small regular giant

(columns A, B, C)

49.
$12M = \begin{matrix} \text{Fairfield} \\ \text{Tyler} \end{matrix} \begin{bmatrix} 2760 & 1080 & 1680 \\ 3120 & 1380 & 1992 \end{bmatrix}$

(columns S.D., N.O., P.M.)

51. $\dfrac{1}{3}\left(\begin{bmatrix} 90 \\ 62 \\ 76 \\ 82 \\ 74 \end{bmatrix} + \begin{bmatrix} 88 \\ 69 \\ 78 \\ 80 \\ 76 \end{bmatrix} + \begin{bmatrix} 91 \\ 73 \\ 72 \\ 84 \\ 77 \end{bmatrix} \right) = \dfrac{1}{3} \begin{bmatrix} 269 \\ 204 \\ 226 \\ 246 \\ 227 \end{bmatrix} = \begin{bmatrix} 89.7 \\ 68.0 \\ 75.3 \\ 82.0 \\ 75.7 \end{bmatrix}$

53. (a) $\begin{bmatrix} 9 & 0 & 8 \\ 6 & 2 & 7 \\ 4 & 8 & 8 \end{bmatrix}$ (b) $\begin{bmatrix} -1.4 & 8.3 \\ 7.7 & 6.6 \end{bmatrix}$ (c) $\begin{bmatrix} 5.4 & 3.0 & 3.1 \\ 3.8 & 4.2 & 1.6 \\ 3.1 & 6.2 & -5.0 \end{bmatrix}$

   (d) $\begin{bmatrix} 3 & 15 & 3 \\ 6 & 3 & 6 \\ 9 & 0 & 12 \end{bmatrix}$ (e) $\begin{bmatrix} 16.20 & 11.88 \\ 32.94 & 39.42 \\ -7.56 & -13.50 \end{bmatrix}$ (f) $\begin{bmatrix} 6 & 9 & 0 \\ 9 & 2 & 13 \\ 0 & 13 & -8 \end{bmatrix}$

Section 2.5

1. $2 + 12 = 14$ 3. $12 + 0 = 12$ 5. $6 + 0 + 8 = 14$

7. $[.90 \quad 1.85 \quad .65] \begin{bmatrix} 2 \\ 1 \\ 4 \end{bmatrix} = 1.80 + 1.85 + 2.60 = \$6.25$

9. $\begin{bmatrix} -5 & 11 \\ 0 & 14 \end{bmatrix}$ 11. $\begin{bmatrix} 30 & 2 \\ 39 & -3 \end{bmatrix}$ 13. $\begin{bmatrix} -7 & 7 \\ -1 & 6 \end{bmatrix}$

15. $\begin{bmatrix} 15 \\ -2 \end{bmatrix}$ 17. Multiplication is not possible.

19. Multiplication is not possible.

21. $\begin{bmatrix} -4 & 4 \\ 3 & 27 \end{bmatrix}$ 23. Multiplication not possible

25. $\begin{bmatrix} 5 & 9 & 12 \\ 5 & 17 & 21 \end{bmatrix}$ 27. $\begin{bmatrix} 8 \\ 13 \\ 7 \end{bmatrix}$

29. $AB = \begin{bmatrix} 1 & 8 \\ -1 & 2 \end{bmatrix}$ $BA = \begin{bmatrix} 3 & 10 \\ -1 & 0 \end{bmatrix}$

31. $AB = \begin{bmatrix} -3 & 10 \\ -2 & 5 \end{bmatrix}$ $BA = \begin{bmatrix} -1 & 2 \\ -4 & 3 \end{bmatrix}$

33. $AB = \begin{bmatrix} 6 & 2 & 13 \\ -11 & -6 & -4 \end{bmatrix}$ BA not possible

35. $AB = [27 \quad 38]$ BA not possible

37. $AB = BA = \begin{bmatrix} 10 & 6 \\ -9 & 10 \end{bmatrix}$

39. $\begin{bmatrix} 10 & -3 \\ 10 & 2 \end{bmatrix} \begin{bmatrix} -1 & 1 \\ 3 & -2 \end{bmatrix} = \begin{bmatrix} -19 & 16 \\ -4 & 6 \end{bmatrix}$

41. $\begin{bmatrix} 8 & 1 & -2 & -14 \\ 5 & 3 & 9 & -11 \\ 8 & 2 & 12 & 2 \end{bmatrix}$ 43. $\begin{bmatrix} 22 & -8 \\ 7 & 1 \end{bmatrix} + \begin{bmatrix} -18 & 12 \\ 3 & 9 \end{bmatrix} = \begin{bmatrix} 4 & 4 \\ 10 & 10 \end{bmatrix}$

45. $\begin{bmatrix} 3 & 4 \\ 1 & 2 \end{bmatrix}$ 47. $\begin{bmatrix} 2 & -10 \\ 3 & 7 \end{bmatrix}$

49. $\begin{bmatrix} 3x + y \\ 2x + 4y \end{bmatrix}$ 51. $\begin{bmatrix} x_1 + 2x_2 - x_3 \\ 3x_1 + x_2 + 4x_3 \\ 2x_1 - x_2 - x_3 \end{bmatrix}$

53. $\begin{bmatrix} x_1 + 3x_2 + 5x_3 + 6x_4 \\ -2x_1 + 9x_2 + 6x_3 + x_4 \\ 8x_1 + 17x_3 + 5x_4 \end{bmatrix}$

55. $\begin{bmatrix} 4 & 5.5 \\ 1 & 2 \end{bmatrix} \begin{bmatrix} 300 \\ 450 \end{bmatrix} = \begin{bmatrix} 3675 \\ 1200 \end{bmatrix}$ 3675 hours assembly time, 1200 hours checking

57. $\begin{bmatrix} 114 & 85 \\ 118 & 84 \\ 116 & 86 \end{bmatrix} \begin{bmatrix} 60 \\ 140 \end{bmatrix} = \begin{bmatrix} 18740 \\ 18840 \\ 19000 \end{bmatrix}$ \$18,740 on Monday, \$18,840 on Wednesday, \$19,000 on Friday

59.
$$\begin{bmatrix} .5 & 1.5 & .5 & 1.0 & 1.0 \\ 0 & 1.0 & 1.0 & 3.0 & 2.0 \end{bmatrix} \begin{bmatrix} 500 & .2 & 0 & 129 \\ 0 & .2 & 0 & 0 \\ 1560 & .32 & 1.7 & 6 \\ 0 & 0 & 0 & 0 \\ 460 & 0 & 0 & 0 \end{bmatrix}$$

$$= \begin{matrix} & A & B_1 & B_2 & C & \\ & \begin{bmatrix} 1490 & .56 & .85 & 67.5 \\ 2480 & .52 & 1.70 & 6 \end{bmatrix} & \begin{matrix} I \\ II \end{matrix} \end{matrix}$$

61.
(a)
$$AB = \begin{matrix} & & Male & Female \\ & Well & \begin{bmatrix} 104750 & 102000 \\ 42000 & 40000 \\ 13250 & 13000 \end{bmatrix} \\ & Sick & \\ & Carrier & \end{matrix}$$

This matrix gives the number of males and the number of females who are well, sick, or carriers.

(b)   42,000 sick males   (c)   102,000 well females

63.   (a)
$$\begin{matrix} & & SEA & LA & DEN & KC & SLC & \\ & SEA & \begin{bmatrix} 0 & 1 & 0 & 0 & 1 \\ 1 & 0 & 1 & 0 & 0 \\ 0 & 1 & 0 & 1 & 1 \\ 0 & 0 & 1 & 0 & 1 \\ 1 & 0 & 1 & 1 & 0 \end{bmatrix} = A \\ & LA & \\ & DEN & \\ & KC & \\ & SLC & \end{matrix}$$

(b)   The desired matrix is A^2.
$$\begin{matrix} & & SEA & LA & DEN & KC & SLC & \\ & SEA & \begin{bmatrix} 2 & 0 & 2 & 1 & 0 \\ 0 & 2 & 0 & 1 & 2 \\ 2 & 0 & 3 & 1 & 1 \\ 1 & 1 & 1 & 2 & 1 \\ 0 & 2 & 1 & 1 & 3 \end{bmatrix} \\ A^2 = & LA & \\ & DEN & \\ & KC & \\ & SLC & \end{matrix}$$

65.   The number of columns of A equals the number of rows of B. The number of rows of A equals the number of columns of B.

67.   BA may not exist, BA may exist and may or may not equal AB.

69.
$$\begin{bmatrix} 1000 \\ 1000 \\ 1000 \end{bmatrix}, \begin{bmatrix} 12000 \\ 500 \\ 250 \end{bmatrix} \begin{bmatrix} 3000 \\ 6000 \\ 125 \end{bmatrix}, \begin{bmatrix} 1500 \\ 1500 \\ 1500 \end{bmatrix}, \begin{bmatrix} 18000 \\ 750 \\ 375 \end{bmatrix}, \begin{bmatrix} 4500 \\ 9000 \\ 187.5 \end{bmatrix}, \begin{bmatrix} 2250 \\ 2250 \\ 2250 \end{bmatrix}$$

The population grows in cycles of three years with the population matrix for a year being equal to 1.5 times the matrix of three years earlier.

71.
$$\begin{bmatrix} 2000 \\ 1000 \\ 1000 \end{bmatrix}, \begin{bmatrix} 4000 \\ 1000 \\ 500 \end{bmatrix}, \begin{bmatrix} 2000 \\ 2000 \\ 500 \end{bmatrix}, \begin{bmatrix} 2000 \\ 1000 \\ 1000 \end{bmatrix}, \ldots$$

The population matrix repeats in cycles of three years.

73.   (a)
$$\begin{bmatrix} 1000 \\ 1000 \\ 1000 \end{bmatrix}, \begin{bmatrix} 3000 \\ 333.3 \\ 500 \end{bmatrix}, \begin{bmatrix} 14000 \\ 10000 \\ 166.67 \end{bmatrix}, \begin{bmatrix} 6400 \\ 4666.67 \\ 5000 \end{bmatrix}, \begin{bmatrix} 148000 \\ 21333.33 \\ 2333.3 \end{bmatrix}, \begin{bmatrix} 184000 \\ 49333.3 \\ 10666.7 \end{bmatrix}$$

The population is increasing indefinitely.

1. $25^{-1} = .04$, $(2/3)^{-1} = 3/2$, $(-5)^{-1} = -1/5$, $(.75)^{-1} = 4/3$,
$11^{-1} = 1/11$

3. $AB = \begin{bmatrix} 1 & 0 & 0 \\ 0 & 1 & 0 \\ 0 & 0 & 1 \end{bmatrix}$ Yes   5. $AB = \begin{bmatrix} 1 & -3 \\ 0 & 10 \end{bmatrix}$ No

7. $AB = \begin{bmatrix} 1 & 0 & 0 \\ 0 & 1 & 0 \\ 0 & 0 & 1 \end{bmatrix}$ Yes

9. $\left[\begin{array}{cc|cc} 1 & 2 & 1 & 0 \\ 3 & 5 & 0 & 1 \end{array} \right]$ $\left[\begin{array}{cc|c} 1 & 2 & 1 & 0 \\ 0 & -1 & -3 & 1 \end{array} \right]$

$\left[\begin{array}{cc|cc} 1 & 0 & -5 & 2 \\ 0 & 1 & 3 & -1 \end{array} \right]$ $A^{-1} = \begin{bmatrix} -5 & 2 \\ 3 & -1 \end{bmatrix}$

11. $\left[\begin{array}{cc|cc} 3 & 2 & 1 & 0 \\ 4 & 3 & 0 & 1 \end{array} \right]$ $\left[\begin{array}{cc|c} 1 & 2/3 & 1/3 & 0 \\ 0 & 1/3 & -4/3 & 1 \end{array} \right]$

$\left[\begin{array}{cc|cc} 1 & 0 & 3 & -2 \\ 0 & 1 & -4 & 1 \end{array} \right]$ $A^{-1} = \begin{bmatrix} 3 & -2 \\ -4 & 3 \end{bmatrix}$

13. $\left[\begin{array}{ccc|ccc} 1 & 3 & 9 & 1 & 0 & 0 \\ 0 & 1 & 4 & 0 & 1 & 0 \\ 3 & 2 & 3 & 0 & 0 & 1 \end{array} \right]$ $\left[\begin{array}{ccc|ccc} 1 & 3 & 9 & 1 & 0 & 0 \\ 0 & 1 & 4 & 0 & 1 & 0 \\ 0 & -7 & -24 & -3 & 0 & 1 \end{array} \right]$

$\left[\begin{array}{ccc|ccc} 1 & 0 & -3 & 1 & -3 & 0 \\ 0 & 1 & 4 & 0 & 1 & 0 \\ 0 & 0 & 4 & -3 & 7 & 1 \end{array} \right]$ $\left[\begin{array}{ccc|ccc} 1 & 0 & 0 & -5/4 & 9/4 & 3/4 \\ 0 & 1 & 0 & 3 & -6 & -1 \\ 0 & 0 & 1 & -3/4 & 7/4 & 1/4 \end{array} \right]$

$A^{-1} = \begin{bmatrix} -5/4 & 9/4 & 3/4 \\ 3 & -6 & -1 \\ -3/4 & 7/4 & 1/4 \end{bmatrix}$

15. $\left[\begin{array}{ccc|ccc} 0 & 4 & -2 & 1 & 0 & 0 \\ 1 & 3 & 5 & 0 & 1 & 0 \\ 1 & 4 & 2 & 0 & 0 & 1 \end{array} \right]$ $\left[\begin{array}{ccc|ccc} 1 & 3 & 5 & 0 & 1 & 0 \\ 0 & 4 & -2 & 1 & 0 & 0 \\ 1 & 4 & 2 & 0 & 0 & 1 \end{array} \right]$

$\left[\begin{array}{ccc|ccc} 1 & 3 & 5 & 0 & 1 & 0 \\ 0 & 4 & -2 & 1 & 0 & 0 \\ 0 & 1 & -3 & 0 & -1 & 1 \end{array} \right]$ $\left[\begin{array}{ccc|ccc} 1 & 0 & 13/2 & -3/4 & 1 & 0 \\ 0 & 1 & -1/2 & 1/4 & 0 & 0 \\ 0 & 0 & -5/2 & -1/4 & -1 & 1 \end{array} \right]$

$\left[\begin{array}{ccc|ccc} 1 & 0 & 0 & -7/5 & -8/5 & 13/5 \\ 0 & 1 & 0 & 3/10 & 1/5 & -1/5 \\ 0 & 0 & 1 & 1/10 & 2/5 & -2/5 \end{array} \right]$ $A^{-1} = \begin{bmatrix} -7/5 & -8/5 & 13/5 \\ 3/10 & 1/5 & -1/5 \\ 1/10 & 2/5 & -2/5 \end{bmatrix}$

17. $\left[\begin{array}{cc|cc} 4 & -2 & 1 & 0 \\ -2 & 1 & 0 & 1 \end{array} \right]$ $\left[\begin{array}{cc|cc} 0 & 0 & 1 & 2 \\ -2 & 1 & 0 & 1 \end{array} \right]$ No inverse

Section 2.6

19. $\begin{bmatrix} 1 & 3 & 1 & \vline & 1 & 0 & 0 \\ 2 & 0 & -2 & \vline & 0 & 1 & 0 \\ 3 & 3 & -1 & \vline & 0 & 0 & 1 \end{bmatrix}$ $\begin{bmatrix} 1 & 3 & 1 & \vline & 1 & 0 & 0 \\ 0 & -6 & -4 & \vline & -2 & 1 & 0 \\ 0 & -6 & -4 & \vline & -3 & 0 & 1 \end{bmatrix}$

$\begin{bmatrix} 1 & 3 & 1 & \vline & 1 & 0 & 0 \\ 0 & -6 & -4 & \vline & -2 & 1 & 0 \\ 0 & 0 & 0 & \vline & -1 & -1 & 1 \end{bmatrix}$ No inverse

21. $\begin{bmatrix} 1 & 2 & 1 & \vline & 1 & 0 & 0 \\ 1 & -3 & 2 & \vline & 0 & 1 & 0 \\ 2 & -1 & 3 & \vline & 0 & 0 & 1 \end{bmatrix}$ $\begin{bmatrix} 1 & 2 & 1 & \vline & 1 & 0 & 0 \\ 0 & -5 & 1 & \vline & -1 & 1 & 0 \\ 0 & -5 & 1 & \vline & -2 & 0 & 1 \end{bmatrix}$

$\begin{bmatrix} 1 & 2 & 1 & \vline & 1 & 0 & 0 \\ 0 & -5 & 1 & \vline & -1 & 1 & 0 \\ 0 & 0 & 0 & \vline & -1 & -1 & 1 \end{bmatrix}$ No inverse

23. $\begin{bmatrix} 2 & 1 & \vline & 1 & 0 \\ 4 & 3 & \vline & 0 & 1 \end{bmatrix}$ $\begin{bmatrix} 1 & 1/2 & \vline & 1/2 & 0 \\ 0 & 1 & \vline & -2 & 1 \end{bmatrix}$ $\begin{bmatrix} 1 & 0 & 3/2 & \vline & -1/2 \\ 0 & 1 & -2 & \vline & 1 \end{bmatrix}$

$A^{-1} = \begin{bmatrix} 3/2 & -1/2 \\ -2 & 1 \end{bmatrix}$

25. $\begin{bmatrix} 1 & 2 & 3 & \vline & 1 & 0 & 0 \\ 2 & -1 & 4 & \vline & 0 & 1 & 0 \\ 0 & -1 & 1 & \vline & 0 & 0 & 1 \end{bmatrix}$ $\begin{bmatrix} 1 & 2 & 3 & \vline & 1 & 0 & 0 \\ 0 & -5 & -2 & \vline & -2 & 1 & 0 \\ 0 & -1 & 1 & \vline & 0 & 0 & 1 \end{bmatrix}$

$\begin{bmatrix} 1 & 2 & 3 & \vline & 1 & 0 & 0 \\ 0 & -1 & 1 & \vline & 0 & 0 & 1 \\ 0 & -5 & -2 & \vline & -2 & 1 & 0 \end{bmatrix}$ $\begin{bmatrix} 1 & 0 & 5 & \vline & 1 & 0 & 2 \\ 0 & 1 & -1 & \vline & 0 & 0 & -1 \\ 0 & 0 & -7 & \vline & -2 & 1 & -5 \end{bmatrix}$

$\begin{bmatrix} 1 & 0 & 0 & \vline & -3/7 & 5/7 & -11/7 \\ 0 & 1 & 0 & \vline & 2/7 & -1/7 & -2/7 \\ 0 & 0 & 1 & \vline & 2/7 & -1/7 & 5/7 \end{bmatrix}$ $A^{-1} = \begin{bmatrix} -3/7 & 5/7 & -11/7 \\ 2/7 & -1/7 & -2/7 \\ 2/7 & -1/7 & 5/7 \end{bmatrix}$

27. (a) $\begin{bmatrix} 3 & 4 & -5 & \vline & 4 \\ 2 & -1 & 3 & \vline & -1 \\ 1 & 1 & -1 & \vline & 2 \end{bmatrix}$ (b) $\begin{bmatrix} 3 & 4 & -5 \\ 2 & -1 & 3 \\ 1 & 1 & -1 \end{bmatrix}$

(c) $\begin{bmatrix} 3 & 4 & -5 \\ 2 & -1 & 3 \\ 1 & 1 & -1 \end{bmatrix} \begin{bmatrix} x_1 \\ x_2 \\ x_3 \end{bmatrix} = \begin{bmatrix} 4 \\ -1 \\ 2 \end{bmatrix}$

29. (a) $\begin{bmatrix} 4 & 5 & \vline & 2 \\ 3 & -2 & \vline & 7 \end{bmatrix}$ (b) $\begin{bmatrix} 4 & 5 \\ 3 & -2 \end{bmatrix}$

(c) $\begin{bmatrix} 4 & 5 \\ 3 & -2 \end{bmatrix} \begin{bmatrix} x \\ y \end{bmatrix} = \begin{bmatrix} 2 \\ 7 \end{bmatrix}$

31. $\begin{bmatrix} 1 & 3 \\ 2 & -1 \end{bmatrix} \begin{bmatrix} x_1 \\ x_2 \end{bmatrix} = \begin{bmatrix} 5 \\ 6 \end{bmatrix}$

33. $\begin{bmatrix} 1 & 2 & -3 & 4 \\ 1 & 1 & 0 & 1 \\ 3 & 2 & 1 & 2 \end{bmatrix} \begin{bmatrix} x_1 \\ x_2 \\ x_3 \\ x_4 \end{bmatrix} = \begin{bmatrix} 0 \\ 5 \\ 4 \end{bmatrix}$

35.
$$\left[\begin{array}{cccc|cccc} 1 & 1 & 0 & 0 & 1 & 0 & 0 & 0 \\ 0 & 1 & 1 & 0 & 0 & 1 & 0 & 0 \\ 1 & 0 & 0 & 1 & 0 & 0 & 1 & 0 \\ 0 & 0 & 1 & 1 & 0 & 0 & 0 & 1 \end{array}\right] \qquad \left[\begin{array}{cccc|cccc} 1 & 1 & 0 & 0 & 1 & 0 & 0 & 0 \\ 0 & 1 & 1 & 0 & 0 & 1 & 0 & 0 \\ 0 & -1 & 0 & 1 & -1 & 0 & 1 & 0 \\ 0 & 0 & 1 & 1 & 0 & 0 & 0 & 1 \end{array}\right]$$

$$\left[\begin{array}{cccc|cccc} 1 & 0 & -1 & 0 & 1 & -1 & 0 & 0 \\ 0 & 1 & 1 & 0 & 0 & 1 & 0 & 0 \\ 0 & 0 & 1 & 1 & -1 & 1 & 1 & 0 \\ 0 & 0 & 1 & 1 & 0 & 0 & 0 & 1 \end{array}\right] \qquad \text{No inverse}$$

37.
$$\left[\begin{array}{ccc|ccc} 1 & 2 & -1 & 1 & 0 & 0 \\ 1 & 1 & 2 & 0 & 1 & 0 \\ 1 & -1 & -1 & 0 & 0 & 1 \end{array}\right] \qquad \left[\begin{array}{ccc|ccc} 1 & 2 & -1 & 1 & 0 & 0 \\ 0 & -1 & 3 & -1 & 1 & 0 \\ 0 & -3 & 0 & -1 & 0 & 1 \end{array}\right]$$

$$\left[\begin{array}{ccc|ccc} 1 & 0 & 5 & -1 & 2 & 0 \\ 0 & 1 & -3 & 1 & -1 & 0 \\ 0 & 0 & -9 & 2 & -3 & 1 \end{array}\right] \qquad \left[\begin{array}{ccc|ccc} 1 & 0 & 5 & -1 & 2 & 0 \\ 0 & 1 & -3 & 1 & -1 & 0 \\ 0 & 0 & 1 & -2/9 & 1/3 & -1/9 \end{array}\right]$$

$$\left[\begin{array}{ccc|ccc} 1 & 0 & 0 & 1/9 & 1/3 & 5/9 \\ 0 & 1 & 0 & 1/3 & 0 & -1/3 \\ 0 & 0 & 1 & -2/9 & 1/3 & -1/9 \end{array}\right] \qquad A^{-1} = \left[\begin{array}{ccc} 1/9 & 1/3 & 5/9 \\ 1/3 & 0 & -1/3 \\ -2/9 & 1/3 & -1/9 \end{array}\right]$$

$$\left[\begin{array}{ccc} 1/9 & 1/3 & 5/9 \\ 1/3 & 0 & -1/3 \\ -2/9 & 1/3 & -1/9 \end{array}\right]\left[\begin{array}{c} 2 \\ 0 \\ 1 \end{array}\right] = \left[\begin{array}{c} 7/9 \\ 1/3 \\ -5/9 \end{array}\right] \qquad x_1 = \frac{7}{9}, \ x_2 = \frac{1}{3}, \ x_3 = \frac{-5}{9}$$

39.
$$\left[\begin{array}{cccc|cccc} 1 & 1 & 2 & 1 & 1 & 0 & 0 & 0 \\ 2 & 0 & -1 & 1 & 0 & 1 & 0 & 0 \\ 0 & 1 & 3 & -1 & 0 & 0 & 1 & 0 \\ 3 & 2 & 0 & 1 & 0 & 0 & 0 & 1 \end{array}\right] \qquad \left[\begin{array}{cccc|cccc} 1 & 1 & 2 & 1 & 1 & 0 & 0 & 0 \\ 0 & -2 & -5 & -1 & -2 & 1 & 0 & 0 \\ 0 & 1 & 3 & -1 & 0 & 0 & 1 & 0 \\ 0 & -1 & -6 & -2 & -3 & 0 & 0 & 1 \end{array}\right]$$

$$\left[\begin{array}{cccc|cccc} 1 & 1 & 2 & 1 & 1 & 0 & 0 & 0 \\ 0 & 1 & 3 & -1 & 0 & 0 & 1 & 0 \\ 0 & -2 & -5 & -1 & -2 & 1 & 0 & 0 \\ 0 & -1 & -6 & -2 & -3 & 0 & 0 & 1 \end{array}\right] \qquad \left[\begin{array}{cccc|cccc} 1 & 0 & -1 & 2 & 1 & 0 & -1 & 0 \\ 0 & 1 & 3 & -1 & 0 & 0 & 1 & 0 \\ 0 & 0 & 1 & -3 & -2 & 1 & 2 & 0 \\ 0 & 0 & -3 & -3 & -3 & 0 & 1 & 1 \end{array}\right]$$

$$\left[\begin{array}{cccc|cccc} 1 & 0 & 0 & -1 & -1 & 1 & 1 & 0 \\ 0 & 1 & 0 & 8 & 6 & -3 & -5 & 0 \\ 0 & 0 & 1 & -3 & -2 & 1 & 2 & 0 \\ 0 & 0 & 0 & -12 & -9 & 3 & 7 & 1 \end{array}\right]$$

$$\left[\begin{array}{cccc|cccc} 1 & 0 & 0 & 0 & -1/4 & 3/4 & 5/12 & -1/12 \\ 0 & 1 & 0 & 0 & 0 & -1 & -1/3 & 2/3 \\ 0 & 0 & 1 & 0 & 1/4 & 1/4 & 1/4 & -1/4 \\ 0 & 0 & 0 & 1 & 3/4 & -1/4 & -7/12 & -1/12 \end{array}\right]$$

$$A^{-1} = \left[\begin{array}{cccc} -1/4 & 3/4 & 5/12 & -1/12 \\ 0 & -1 & -1/3 & 2/3 \\ 1/4 & 1/4 & 1/4 & -1/4 \\ 3/4 & -1/4 & -7/12 & -1/12 \end{array}\right] \qquad A^{-1}\left[\begin{array}{c} 4 \\ 6 \\ 3 \\ 9 \end{array}\right] = \left[\begin{array}{c} 4 \\ -1 \\ 1 \\ -1 \end{array}\right]$$

$$x_1 = 4, \ x_2 = -1, \ x_3 = 1, \ x_4 = -1$$

41. $\begin{bmatrix} -2 & 1 & 3 & | & 1 & 0 & 0 \\ 2 & 4 & -1 & | & 0 & 1 & 0 \\ 3 & 0 & -4 & | & 0 & 0 & 1 \end{bmatrix}$ $\begin{bmatrix} 1 & -1/2 & -3/2 & | & -1/2 & 0 & 0 \\ 0 & 5 & 2 & | & 1 & 1 & 0 \\ 0 & 3/2 & 1/2 & | & 3/2 & 0 & 1 \end{bmatrix}$

$\begin{bmatrix} 1 & 0 & -13/10 & | & -2/5 & 1/10 & 0 \\ 0 & 1 & 2/5 & | & 1/5 & 1/5 & 0 \\ 0 & 0 & -1/10 & | & 6/5 & -3/10 & 1 \end{bmatrix}$ $\begin{bmatrix} 1 & 0 & 0 & | & -16 & 4 & -13 \\ 0 & 1 & 0 & | & 5 & -1 & 4 \\ 0 & 0 & 1 & | & -12 & 3 & -10 \end{bmatrix}$

$A^{-1} = \begin{bmatrix} -16 & 4 & -13 \\ 5 & -1 & 4 \\ -12 & 3 & -10 \end{bmatrix}$ $A^{-1}\begin{bmatrix} 1 \\ 5 \\ 2 \end{bmatrix} = \begin{bmatrix} -22 \\ 8 \\ -17 \end{bmatrix}$ $(-22, 8, -17)$

$A^{-1}\begin{bmatrix} -1 \\ 3 \\ 1 \end{bmatrix} = \begin{bmatrix} 15 \\ -4 \\ 11 \end{bmatrix}$ $x_1 = 15, \; x_2 = -4, \; x_3 = 11$

$A^{-1}\begin{bmatrix} 0 \\ 1 \\ 2 \end{bmatrix} = \begin{bmatrix} -22 \\ 7 \\ -17 \end{bmatrix}$ $x_1 = -22, \; x_2 = 7, \; x_3 = -17$

43. $\begin{bmatrix} 1 & 2 & | & 1 & 0 \\ 3 & 5 & | & 0 & 1 \end{bmatrix}$ $\begin{bmatrix} 1 & 2 & | & 1 & 0 \\ 0 & -1 & | & -3 & 1 \end{bmatrix}$

$\begin{bmatrix} 1 & 0 & | & -5 & 2 \\ 0 & 1 & | & 3 & -1 \end{bmatrix}$ $A^{-1} = \begin{bmatrix} -5 & 2 \\ 3 & -1 \end{bmatrix}$

(a) $A^{-1}\begin{bmatrix} 3 \\ 8 \end{bmatrix} = \begin{bmatrix} 1 \\ 1 \end{bmatrix}$ $x_1 = 1, \; x_2 = 1$

(b) $A^{-1}\begin{bmatrix} 4 \\ 9 \end{bmatrix} = \begin{bmatrix} -2 \\ 3 \end{bmatrix}$ $x_1 = -2, \; x_2 = 3$

(c) $A^{-1}\begin{bmatrix} 3 \\ 7 \end{bmatrix} = \begin{bmatrix} -1 \\ 2 \end{bmatrix}$ $x_1 = -1, \; x_2 = 2$

45. (a) Vitamin C intake = 32x + 24y, Vitamin A intake = 900x + 425y

so $\begin{bmatrix} 32 & 24 \\ 900 & 425 \end{bmatrix} \begin{bmatrix} x \\ y \end{bmatrix} = \begin{bmatrix} b_1 \\ b_2 \end{bmatrix}$

where b_1 = vitamin C intake and b_2 = vitamin A intake

(b) $\begin{bmatrix} 32 & 24 \\ 900 & 425 \end{bmatrix} \begin{bmatrix} 3.2 \\ 2.5 \end{bmatrix} = \begin{bmatrix} 162.4 \\ 3942.5 \end{bmatrix}$

162.4 mg of C, 3,942.5 iu of A

(c) $\begin{bmatrix} 32 & 24 \\ 900 & 425 \end{bmatrix} \begin{bmatrix} 1.5 \\ 3.0 \end{bmatrix} = \begin{bmatrix} 120 \\ 2625 \end{bmatrix}$

120 mg of C, 2,625 iu of A

(d) $\begin{bmatrix} -.053125 & .003 \\ .1125 & -.004 \end{bmatrix} \begin{bmatrix} 107.2 \\ 2315.0 \end{bmatrix} = \begin{bmatrix} 1.25 \\ 2.8 \end{bmatrix}$

1.25 units of A, 2.8 units of B

(e) $\begin{bmatrix} -.053125 & .003 \\ .1125 & -.004 \end{bmatrix} \begin{bmatrix} 104 \\ 2575 \end{bmatrix} = \begin{bmatrix} 2.2 \\ 1.4 \end{bmatrix}$

2.2 units of A, 1.4 units of B

47. (a) $\begin{bmatrix} 2 & -.25 \\ -1 & .25 \end{bmatrix} \begin{bmatrix} 900 \\ 5840 \end{bmatrix} = \begin{bmatrix} 340 \\ 560 \end{bmatrix}$

340 children, 560 adults

(b) $\begin{bmatrix} 2 & -.25 \\ -1 & .25 \end{bmatrix} \begin{bmatrix} 1000 \\ 6260 \end{bmatrix} = \begin{bmatrix} 435 \\ 565 \end{bmatrix}$

435 children, 565 adults

(c) $\begin{bmatrix} 2 & -.25 \\ -1 & .25 \end{bmatrix} \begin{bmatrix} 750 \\ 5560 \end{bmatrix} = \begin{bmatrix} 110 \\ 640 \end{bmatrix}$

110 children, 640 adults

49. $A^{-1} = \begin{bmatrix} 1 & -1 & 0 \\ -2 & 5 & 1 \\ -1 & 3 & 1 \end{bmatrix}$

51. $\begin{bmatrix} .4 & .8 & -.2 & -.8 \\ -.4 & .2 & .2 & -.2 \\ 0 & -.1429 & .1429 & 0 \\ -.1 & -.4857 & .0857 & .7 \end{bmatrix}$

53. (a) $AB = \begin{bmatrix} 1 & 0 & 1 \\ 0 & 0 & 1 \\ 1 & -1 & 1 \end{bmatrix}$ $A^{-1} = \begin{bmatrix} 1 & -1 & 0 \\ 0 & 1 & -1 \\ 0 & 0 & 1 \end{bmatrix}$ $B^{-1} = \begin{bmatrix} 1 & 0 & 0 \\ 1 & 1 & 0 \\ 0 & 1 & 1 \end{bmatrix}$

$(AB)^{-1} = \begin{bmatrix} 1 & -1 & 0 \\ 1 & 0 & -1 \\ 0 & 1 & 0 \end{bmatrix}$ $A^{-1}B^{-1} = \begin{bmatrix} 0 & -1 & 0 \\ 1 & 0 & -1 \\ 0 & 1 & 1 \end{bmatrix}$

$B^{-1}A^{-1} = \begin{bmatrix} 1 & -1 & 0 \\ 1 & 0 & -1 \\ 0 & 1 & 0 \end{bmatrix}$

(b) $(AB)^{-1} = B^{-1}A^{-1}$

55. (a) $AB = \begin{bmatrix} 9 & 14 \\ 5 & 8 \end{bmatrix}$ $A^{-1} = \begin{bmatrix} 2 & -3 \\ -1 & 2 \end{bmatrix}$ $B^{-1} = \begin{bmatrix} 1 & -2 \\ -.5 & 1.5 \end{bmatrix}$

$(AB)^{-1} = \begin{bmatrix} 4 & -7 \\ -2.5 & 4.5 \end{bmatrix}$ $A^{-1}B^{-1} = \begin{bmatrix} 3.5 & -8.5 \\ -2 & 5 \end{bmatrix}$

$B^{-1}A^{-1} = \begin{bmatrix} 4 & -7 \\ -2.5 & 4.5 \end{bmatrix}$

(b) $(AB)^{-1} = B^{-1}A^{-1}$

57. (a) $A^{-1} = \begin{bmatrix} 2 & -1.5 & 2.5 \\ .5 & -.25 & .75 \\ 0 & -.25 & .25 \end{bmatrix}$

The reduced echelon form of A is $\begin{bmatrix} 1 & 0 & 0 \\ 0 & 1 & 0 \\ 0 & 0 & 1 \end{bmatrix}$.

(b) The graphing calculator gives an error when trying to compute A^{-1} indicating that no inverse exists. The reduced echelon form of A is $\begin{bmatrix} 1 & 0 & 2 \\ 0 & 1 & 0 \\ 0 & 0 & 0 \end{bmatrix}$.

59. When the reduced form of A is the identity matrix, A^{-1} exists. When the reduced form of A contains a row of zeros, A^{-1} does not exist.

61. $\begin{bmatrix} 107 \\ 12 \\ -44 \end{bmatrix}$

Section 2.7

1. $AX = \begin{bmatrix} .15 & .08 \\ .30 & .20 \end{bmatrix} \begin{bmatrix} 8 \\ 12 \end{bmatrix} = \begin{bmatrix} 2.16 \\ 4.8 \end{bmatrix}$

3. $\begin{bmatrix} .06 & .12 & .09 \\ .15 & .05 & .10 \\ .08 & .04 & .02 \end{bmatrix} \begin{bmatrix} 8 \\ 14 \\ 10 \end{bmatrix} = \begin{bmatrix} 3.06 \\ 2.90 \\ 1.40 \end{bmatrix}$

5. $\left[\begin{array}{cc|cc} .8 & -.3 & 1 & 0 \\ -.2 & .7 & 0 & 1 \end{array} \right]$ $\left[\begin{array}{cc|cc} 1 & -.375 & 1.25 & 0 \\ 0 & .625 & .25 & 1 \end{array} \right]$

    $\left[\begin{array}{cc|cc} 1 & 0 & 1.4 & .6 \\ 0 & 1 & .4 & 1.6 \end{array} \right]$ $(I - A)^{-1} = \begin{bmatrix} 1.4 & .6 \\ .4 & 1.6 \end{bmatrix}$

7. Find $(I-A)^{-1}D$

    $\left[\begin{array}{cc|cc} .76 & -.08 & 1 & 0 \\ -.12 & .96 & 0 & 1 \end{array} \right]$ $\left[\begin{array}{cc|cc} 1 & -.1053 & 1.3158 & 0 \\ 0 & .9474 & 0.15789 & 1 \end{array} \right]$

    $\left[\begin{array}{cc|cc} 1 & 0 & 1.333 & 0.1111 \\ 0 & 1 & 0.1667 & 1.0555 \end{array} \right]$

    $(I - A)^{-1}D = \begin{bmatrix} 1.3333 & 0.1111 \\ 0.1667 & 1.0555 \end{bmatrix} \begin{bmatrix} 15 \\ 12 \end{bmatrix} = \begin{bmatrix} 21.33 \\ 15.17 \end{bmatrix}$

9. $I - A = \begin{bmatrix} .8 & -.4 \\ -.4 & .7 \end{bmatrix}$ and $(I - A)^{-1} = \begin{bmatrix} 1.75 & 1 \\ 1 & 2 \end{bmatrix}$

    The output levels required to meet the demands $\begin{bmatrix} 20 \\ 28 \end{bmatrix}$ and $\begin{bmatrix} 15 \\ 11 \end{bmatrix}$

    are obtained from $X = \begin{bmatrix} 1.75 & 1 \\ 1 & 2 \end{bmatrix} \begin{bmatrix} 20 & 15 \\ 28 & 11 \end{bmatrix} = \begin{bmatrix} 63 & 37.25 \\ 76 & 37.00 \end{bmatrix}$

    and are $\begin{bmatrix} 63 \\ 76 \end{bmatrix}$ and $\begin{bmatrix} 37.25 \\ 37.00 \end{bmatrix}$, respectively.

11. $I - A = \begin{bmatrix} .8 & -.2 & -.2 \\ -.1 & .4 & -.2 \\ -.1 & -.1 & .6 \end{bmatrix}$ and $(I - A)^{-1} = \begin{bmatrix} \frac{22}{15} & \frac{14}{15} & \frac{4}{5} \\ \frac{8}{15} & \frac{46}{15} & \frac{6}{5} \\ \frac{1}{3} & \frac{2}{3} & 2 \end{bmatrix}$

    The values of X required to meet the demands $\begin{bmatrix} 30 \\ 24 \\ 42 \end{bmatrix}$ and $\begin{bmatrix} 60 \\ 45 \\ 75 \end{bmatrix}$ come

    from the solution of

$$X = \begin{bmatrix} \frac{22}{15} & \frac{14}{15} & \frac{4}{5} \\ \frac{8}{15} & \frac{46}{15} & \frac{6}{5} \\ \frac{1}{3} & \frac{2}{3} & 2 \end{bmatrix} \begin{bmatrix} 30 & 60 \\ 24 & 45 \\ 42 & 75 \end{bmatrix} = \begin{bmatrix} 100 & 190 \\ 140 & 260 \\ 110 & 200 \end{bmatrix}$$

and are $\begin{bmatrix} 100 \\ 140 \\ 110 \end{bmatrix}$ and $\begin{bmatrix} 190 \\ 260 \\ 200 \end{bmatrix}$, respectively.

13. (a) $AX = \begin{bmatrix} 23.5 \\ 28.5 \end{bmatrix}$ (b) $X - AX = \begin{bmatrix} 16.5 \\ 21.5 \end{bmatrix}$

15. (a) $AX = \begin{bmatrix} 21.6 \\ 64.8 \\ 36.0 \end{bmatrix}$ (b) $X - AX = \begin{bmatrix} 14.4 \\ 7.2 \\ 0 \end{bmatrix}$

17. (a) The input-output matrix is

$$A = \begin{matrix} A \\ N \end{matrix} \begin{matrix} A & N \\ \begin{bmatrix} .1 & .3 \\ .6 & .4 \end{bmatrix} \end{matrix}$$

(b) Internal consumption = $\begin{bmatrix} .1 & .3 \\ .6 & .4 \end{bmatrix} \begin{bmatrix} 3.5 \\ 5.2 \end{bmatrix} = \begin{bmatrix} 1.91 \\ 4.18 \end{bmatrix}$
Agriculture internal consumption = \$1.91 million leaving
\$1.59 million for export.
Nonagriculture consumption = \$4.18 million leaving \$1.02
million for export.

(c) $D = \begin{bmatrix} 2 \\ 2 \end{bmatrix}$. The total production required is given by X
where $(I - A)^{-1} D = X$.
$I - A = \begin{bmatrix} .9 & -.3 \\ -.6 & .6 \end{bmatrix}$ and $(I - A)^{-1} = \begin{bmatrix} 5/3 & 5/6 \\ 5/3 & 5/2 \end{bmatrix}$

Total production = $\begin{bmatrix} 5/3 & 5/6 \\ 5/3 & 5/2 \end{bmatrix} \begin{bmatrix} 2 \\ 2 \end{bmatrix} = \begin{bmatrix} 5 \\ 8.3 \end{bmatrix}$
To export \$2 million of each kind of product requires
production of \$5 million of agriculture and \$8.3 million of
nonagriculture products.

(d) For \$2 million of agriculture and \$3 million of
nonagriculture exports
Total production = $\begin{bmatrix} 5/3 & 5/6 \\ 5/3 & 5/2 \end{bmatrix} \begin{bmatrix} 2 \\ 3 \end{bmatrix} = \begin{bmatrix} 5.83 \\ 10.83 \end{bmatrix}$
Thus, \$5.83 of agriculture and \$10.83 million of
nonagriculture products are required.

19.

(a) $A = \begin{matrix} P \\ E \end{matrix} \begin{matrix} P & E \\ \begin{bmatrix} .10 & .40 \\ .20 & .20 \end{bmatrix} \end{matrix}$

(b) The amounts used internally are $\begin{bmatrix} .10 & .40 \\ .20 & .20 \end{bmatrix} \begin{bmatrix} 25 \\ 32 \end{bmatrix} = \begin{bmatrix} 15.3 \\ 11.4 \end{bmatrix}$
\$15.3 million worth of plastics and \$11.4 million worth of
electronics are used internally.

(c) Solve $(I - A)^{-1} \begin{bmatrix} 36 \\ 44 \end{bmatrix} = X$ where X represents total
production.

43

$$I - A = \begin{bmatrix} .90 & -.40 \\ -.20 & .80 \end{bmatrix} \qquad (I - A)^{-1} = \begin{bmatrix} 1.25 & .625 \\ .3125 & 1.40625 \end{bmatrix}$$

$$(I - A)^{-1} \begin{bmatrix} 36 \\ 44 \end{bmatrix} = \begin{bmatrix} 72.5 \\ 73.125 \end{bmatrix}$$

The corporation must produce $72.5 million worth of plastics and $73.125 million worth of electronics to have $36 million worth of plastics and $44 million worth of electronics available for external sales.

21.

(a)
$$\begin{array}{c} \\ C \\ A = M \\ US \end{array} \begin{array}{ccc} C & M & US \\ \begin{bmatrix} .2 & .1 & .3 \\ .2 & .4 & 0 \\ .4 & 0 & .3 \end{bmatrix} \end{array}$$

(b) The value of components used internally is

$$\begin{bmatrix} .2 & .1 & .3 \\ .2 & .4 & 0 \\ .4 & 0 & .3 \end{bmatrix} \begin{bmatrix} 10 \\ 18 \\ 15 \end{bmatrix} = \begin{bmatrix} 8.3 \\ 9.2 \\ 8.5 \end{bmatrix}$$

Canada uses $8.3 million worth of components, Mexico $9.2 million, and United States $8.5 million.

(c) Solve $(I - A)^{-1} \begin{bmatrix} 24 \\ 30 \\ 20 \end{bmatrix}$

$$I - A = \begin{bmatrix} .8 & -.1 & -.3 \\ -.2 & .6 & 0 \\ -.4 & 0 & .7 \end{bmatrix}$$

$$(I - A)^{-1} = \begin{bmatrix} 1.68 & .28 & .72 \\ .56 & 1.76 & .24 \\ .96 & .16 & 1.84 \end{bmatrix} \text{ (rounded to 2 decimals)}$$

$$(I - A)^{-1} \begin{bmatrix} 24 \\ 30 \\ 20 \end{bmatrix} = \begin{bmatrix} 63.12 \\ 71.04 \\ 64.64 \end{bmatrix}$$

Canada must produce $63.12 million worth of vehicles, Mexico $71.04 million, and United States $64.64 million.

23. A negative entry indicates that a negative cost is associated with producing a good.
An entry greater than 1 indicates it costs more than $1.00 to produce $1.00 worth of a good.

25. $\begin{bmatrix} 1.233 & .222 & .164 \\ .656 & 1.578 & .688 \\ .889 & .847 & 1.578 \end{bmatrix}$

27. $I - A = \begin{bmatrix} .60 & -.10 & -.20 & -.15 \\ -.20 & .80 & -.15 & -.10 \\ -.02 & -.03 & .99 & -.12 \\ -.25 & -.04 & -.01 & .90 \end{bmatrix}$

$$(I - A)^{-1} = \begin{bmatrix} 1.948 & .281 & .440 & .414 \\ .582 & 1.349 & .325 & .290 \\ .126 & .063 & 1.047 & .168 \\ .568 & .139 & .148 & 1.241 \end{bmatrix}$$

$$(I - A)^{-1}\begin{bmatrix}450,000\\300,000\\620,000\\240,000\end{bmatrix} = \begin{bmatrix}1,333,000\\937,550\\764,890\\687,110\end{bmatrix}$$ production for year 1

$$(I - A)^{-1}\begin{bmatrix}500,000\\325,000\\600,000\\250,000\end{bmatrix} = \begin{bmatrix}1,432,740\\996,770\\753,510\\728,440\end{bmatrix}$$ production for year 2

$$(I - A)^{-1}\begin{bmatrix}475,000\\360,000\\590,000\\280,000\end{bmatrix} = \begin{bmatrix}1,401,910\\1,034,920\\747,140\\754,830\end{bmatrix}$$ production for year 3

Review Exercises, Chapter 2

1. $3x + 2y = 5$
    $2x + 4y = 9$

    $x = 9/2 - 2y$
    $3(9/2 - 2y) + 2y = 5$
    $-4y = 5 - 27/2$
    $4y = 17/2$ so $y = 17/8$
    $x = 9/2 - 2(17/8)$
     $= 9/2 - 17/4 = 1/4$
    $(1/4, 17/8)$

3. $5x - y = 34$
    $2x + 3y = 0$

    $15x - 3y = 102$
    $\underline{2x + 3y = 0}$
    $17x = 102$
     $x = 6$
     $y = -4$

5. $x - 2y + 3z = 3$
    $4x + 7y - 6z = 6$
    $-2x + 4y + 12z = 0$

    Use the first and second equations
         $4x - 8y + 12z = 12$
         $\underline{4x + 7y - 6z = 6}$
          $- 15y + 18z = 6$
    Use the first and third equations
         $2x - 4y + 6z = 6$
         $\underline{-2x + 4y + 12z = 0}$
              $18z = 6$ so $z = 1/3$
         $-15y + 6 = 6$ gives $y = 0$ which then gives $x = 2$ so the solution is $(2, 0, 1/3)$

7. $\begin{bmatrix}2 & -4 & -14 & | & 50\\1 & -1 & -5 & | & 17\\2 & -4 & -17 & | & 65\end{bmatrix}$ $\begin{bmatrix}1 & -1 & -5 & | & 17\\2 & -4 & -14 & | & 50\\2 & -4 & -17 & | & 65\end{bmatrix}$ $\begin{bmatrix}1 & -1 & -5 & | & 17\\0 & -2 & -4 & | & 16\\0 & -2 & -7 & | & 31\end{bmatrix}$

    $\begin{bmatrix}1 & 0 & -3 & | & 9\\0 & 1 & 2 & | & -8\\0 & 0 & -3 & | & 15\end{bmatrix}$ $\begin{bmatrix}1 & 0 & 0 & | & -6\\0 & 1 & 0 & | & 2\\0 & 0 & 1 & | & -5\end{bmatrix}$ $x_1 = -6, x_2 = 2, x_3 = -5$

9. $\begin{bmatrix}1 & -1 & | & 3\\4 & 3 & | & 5\\6 & 1 & | & 9\end{bmatrix}$ $\begin{bmatrix}1 & -1 & | & 3\\0 & 7 & | & -7\\0 & 7 & | & -9\end{bmatrix}$ $\begin{bmatrix}1 & -1 & | & 3\\0 & 7 & | & -7\\0 & 0 & | & -2\end{bmatrix}$
    No solution

11.
$$\begin{bmatrix} 1 & 0 & 1 & 0 \\ 2 & -1 & 1 & -1 \\ 1 & -1 & 0 & -1 \end{bmatrix} \quad \begin{bmatrix} 1 & 0 & 1 & 0 \\ 0 & -1 & -1 & -1 \\ 0 & -1 & -1 & -1 \end{bmatrix} \quad \begin{bmatrix} 1 & 0 & 1 & 0 \\ 0 & 1 & 1 & 1 \\ 0 & 0 & 0 & 0 \end{bmatrix}$$

$x = -z, \ y = 1 - z$

13.
$$\begin{bmatrix} 1 & 2 & -1 & 3 & 3 \\ 1 & 3 & 1 & -1 & 0 \\ 2 & 1 & -6 & 2 & -11 \\ 3 & 7 & -1 & 5 & 6 \end{bmatrix} \quad \begin{bmatrix} 1 & 2 & -1 & 3 & 3 \\ 0 & 1 & 2 & -4 & -3 \\ 0 & -3 & -4 & -4 & -17 \\ 0 & 1 & 2 & -4 & -3 \end{bmatrix}$$

$$\begin{bmatrix} 1 & 0 & -5 & 11 & 9 \\ 0 & 1 & 2 & -4 & -3 \\ 0 & 0 & 2 & -16 & -26 \\ 0 & 0 & 0 & 0 & 0 \end{bmatrix} \quad \begin{bmatrix} 1 & 0 & 0 & -29 & -56 \\ 0 & 1 & 0 & 12 & 23 \\ 0 & 0 & 1 & -8 & -13 \\ 0 & 0 & 0 & 0 & 0 \end{bmatrix}$$

$x_1 = -56 + 29x_4, \ x_2 = 23 - 12x_4, \ x_3 = -13 + 8x_4$

15.
$$\begin{bmatrix} 2 & 3 & -5 & 8 \\ 6 & -3 & 1 & 16 \end{bmatrix} \quad \begin{bmatrix} 1 & 3/2 & -5/2 & 4 \\ 0 & -12 & 16 & -8 \end{bmatrix}$$

$$\begin{bmatrix} 1 & 0 & -1/2 & 3 \\ 0 & 1 & -4/3 & 2/3 \end{bmatrix} \quad x_1 = 3 + (1/2)x_3, \ x_2 = 2/3 + (4/3)x_3$$

17. $3x + 2 = 5 - x$
$4x = 3$
$x = 3/4$

19.
$$\begin{bmatrix} -3 & -2 \\ 6 & 7 \end{bmatrix}$$

21.
$$\begin{bmatrix} 11 & -3 \\ 7 & -1 \\ 3 & 0 \end{bmatrix}$$

23.   [3]

25.   Cannot multiply them

27.
$$\begin{bmatrix} 8 & 6 & 1 & 0 \\ 7 & 5 & 0 & 1 \end{bmatrix} \quad \begin{bmatrix} 1 & 3/4 & 1/8 & 0 \\ 0 & -1/4 & -7/8 & 1 \end{bmatrix}$$

$$\begin{bmatrix} 1 & 0 & -5/2 & 3 \\ 0 & 1 & 7/2 & -4 \end{bmatrix} \quad A^{-1} = \begin{bmatrix} -5/2 & 3 \\ 7/2 & -4 \end{bmatrix}$$

29.
$$\begin{bmatrix} 1 & 0 & 3 & 1 & 0 & 0 \\ 2 & -5 & 4 & 0 & 1 & 0 \\ 1 & -2 & 2 & 0 & 0 & 1 \end{bmatrix} \quad \begin{bmatrix} 1 & 0 & 3 & 1 & 0 & 0 \\ 0 & -5 & -2 & -2 & 1 & 0 \\ 0 & -2 & -1 & -1 & 0 & 1 \end{bmatrix}$$

$$\begin{bmatrix} 1 & 0 & 3 & 1 & 0 & 0 \\ 0 & 1 & 2/5 & 2/5 & -1/5 & 0 \\ 0 & 0 & -1/5 & -1/5 & -2/5 & 1 \end{bmatrix} \quad \begin{bmatrix} 1 & 0 & 0 & -2 & -6 & 15 \\ 0 & 1 & 0 & 0 & -1 & 2 \\ 0 & 0 & 1 & 1 & 2 & -5 \end{bmatrix}$$

$$A^{-1} = \begin{bmatrix} -2 & -6 & 15 \\ 0 & -1 & 2 \\ 1 & 2 & -5 \end{bmatrix}$$

31.
$$\begin{bmatrix} 6 & 4 & -5 & 10 \\ 3 & -2 & 0 & 12 \\ 1 & 1 & -4 & -2 \end{bmatrix}$$

33. $\begin{bmatrix} 2 & 4 & 6 & -2 \\ 3 & 1 & 0 & 5 \\ -2 & 1 & 3 & -11 \end{bmatrix}$ $\begin{bmatrix} 1 & 2 & 3 & -1 \\ 0 & -5 & -9 & 8 \\ 0 & 5 & 9 & -13 \end{bmatrix}$

$\begin{bmatrix} 1 & 0 & -3/5 & 11/5 \\ 0 & 1 & 9/5 & -8/5 \\ 0 & 0 & 0 & -5 \end{bmatrix}$ $\begin{bmatrix} 1 & 0 & -3/5 & 0 \\ 0 & 1 & 9/5 & 0 \\ 0 & 0 & 0 & 1 \end{bmatrix}$

35. Let x = number of field goals, y = number of free throws.

$x + y = 36$
$2x + y = 59$

$\left[\begin{array}{cc|c} 1 & 1 & 36 \\ 2 & 1 & 59 \end{array} \right]$ $\left[\begin{array}{cc|c} 1 & 1 & 36 \\ 0 & -1 & -13 \end{array} \right]$ $\left[\begin{array}{cc|c} 1 & 0 & 23 \\ 0 & 1 & 13 \end{array} \right]$

23 field goals, 13 free throws

37. Let x = amount invested in bonds, y = amount invested in stocks.

$x + y = 50,000$
$.07x + .12y = 5,000$

$\left[\begin{array}{cc|c} 1 & 1 & 50,000 \\ .07 & .12 & 5,000 \end{array} \right]$ $\left[\begin{array}{cc|c} 1 & 1 & 50,000 \\ 0 & .05 & 1,500 \end{array} \right]$ $\left[\begin{array}{cc|c} 1 & 0 & 20,000 \\ 0 & 1 & 30,000 \end{array} \right]$

$20,000 in bonds, $30,000 in stocks

39. Let x = shares of High-Tech, y = shares of Big Burger

$38x + 16y = 5648$
$40.5x + 15.75y = 5931$

$\left[\begin{array}{cc|c} 1 & 8/19 & 2824/19 \\ 0 & -99/76 & -1683/19 \end{array} \right]$ $\left[\begin{array}{cc|c} 1 & 0 & 120 \\ 0 & 1 & 68 \end{array} \right]$

120 of High-Tech, 68 of Big Burger

41. Let x = number at plant A, y = number at plant B.

$x + y = 900$
$3.6x + 1260 = 3.3y + 2637$
$3.6x + 1260 = 3.3(900 - x) + 2637$
$6.9x = 4347$
$x = 630, y = 270$ so   630 at plant A and 270 at plant B.

43. $P(-6) = 4600, P(0) = 5400$ so the slope of the line is

$$m = \frac{5400 - 4600}{0 - (-6)} = \frac{400}{3}$$

The y-intercept is 5400 so the equation is $P(t) = \frac{400}{3}t + 5400$

$P(15) = \frac{400}{3}(15) + 5400 = 2000 + 5400 = 7400$

Chapter 3
Linear Programming

Section 3.1

1.   (1, -1):   5(1) + 2(-1) = 5 - 2 = 3   17   yes
       (4, 1) :   5(4) + 2(1) = 20 + 2 | 17   no
       (3, 1) :   5(3) + 2(1) = 15 + 2 = 17   17   yes
       (4, 4) :   5(4) + 2(4) = 20 + 8 = 28 | 17   no
       (2, 3) :   5(2) + 2(3) = 10 + 6 = 16   17   yes

3.   6x + 8y   24
Find the x and y intercepts
x = 0: 8y = 24   y = 3
y = 0: 6x = 24   x = 4

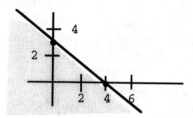

5.   3x - 7y   21
Find the x and y intercepts
x = 0: -7y = 21   y = -3
y = 0: 3x = 21   x = 7

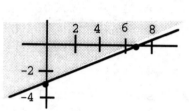

7.   5x + 4y < 20
Find the x and y intercepts
x = 0: 4y = 20   y = 5
y = 0: 5x = 20   x = 4

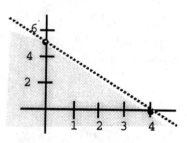

9.   6x + 5y < 30
Find the x and y intercepts
x = 0: 5y = 30   y = 6
y = 0: 6x = 30   x = 5

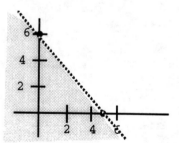

11.   x   10

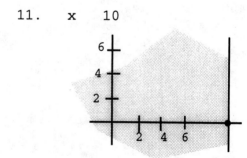

13.   y ≥ -3

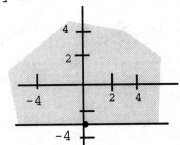

15.   9x - 6y > 30
       Find the x and y intercepts
       x = 0: -6y = 30   y = -5
       y = 0: 9x = 30   x = 10/3

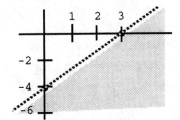

17.   4x - 3y > 12
       Find the x and y intercepts
       x = 0: -3y = 12   y = -4
       y = 0: 4x = 12   x = 3

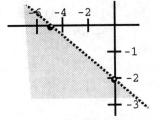

19.   -2x - 5y > 10
       Find the x and y intercepts
       x = 0: -5y = 10   y = -2
       y = 0: -2x = 10   x = -5

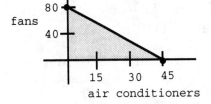

21.   Let x = number of air conditioners,
       y = number of fans.
       3.2x + 1.8y ≤ 144
       Find the x and y intercepts
       x = 0: 1.8y = 144 y = 80
       y = 0: 3.2x = 144 x = 45

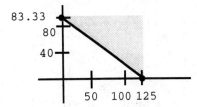

23.   Let x = number of members,
       y = number of pledges
       (a)   4x + 6y ≥ 500
       (b)   x = 0: 6y = 500   y = 83.33
              y = 0: 4x = 500   x = 125

25.   Let x = number of TV spots,
       y = number of newspaper ads.
            900x + 830y ≤ 75,000

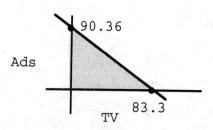

27.   Let x = number of acres of strawberries, y = number of acres of
       tomatoes.   9x + 6y ≤ 750

29.   Let x = number of days the Glen Echo plant operates, y = number of
       days the Speegleville Road plant operates.
              (a)   200x + 300y ≥ 2400   (paperbacks)
              (b)   300x + 200y ≥ 2100   (hardbacks)

31.   Let x = number of drills and y = number of screwdrivers. The
       constraints are
              2x + 4y ≥ 900   (Production time)
               x + y ≤ 300   (Packing capacity)
                    y ≥ 2x   (More screwdrivers than drills)
            x ≥ 0, y ≥ 0
       Maximize   z = 40x + 30y   (income)

33.   Let x represent the number of servings of milk. Let y represent
       the number of servings of bread. 12x + 15y ≥ 50.

Section 3.2

1.   x + y ≤ 3   Intercepts: (3,0),(0,3)
       2x - y < -2   Intercepts: (-1,0),(0,2)

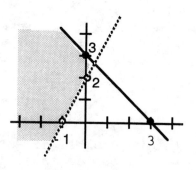

3.   x ≥ 3
       y ≥ 2
       3x + 2y < 18   Intercepts: (6,0),(0,9)

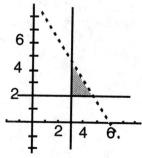

5. $4x + 6y \leq 18$ Intercepts: $(9/2,0),(0,3)$
       $x + 3y \leq 6$ Intercepts: $(6,0),(0,2)$
       $x \geq 0$

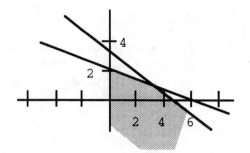

7. $2x + y \leq 60$ Intercepts: $(30,0), (0,60)$
      $2x + 3y \leq 120$ Intercepts: $(60,0), (0,40)$

      $x \geq 0, y \geq 0$

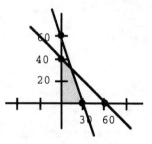

9. $x + y > 4$ Intercepts: $(4,0), (0,4)$
      $2x - 3y \geq 8$ Intercepts: $(4,0), (0,-8/3)$

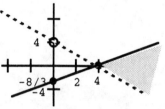

Corner: $(4,0)$

11. $-3x + 10y = 15$ Intercepts: $(-5, 0)$ and $(0, 1.5)$
       $3x + 5y = 15$ Intercepts: $(5, 0)$ and $(0, 3)$

The lines intersect at $(\frac{5}{3}, 2)$.

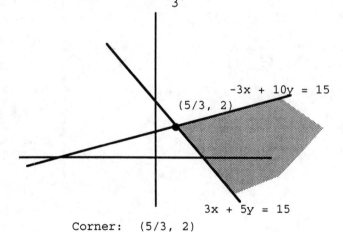

Corner: $(5/3, 2)$

13. $x \leq 0$
    $y \geq 2$
    The y-axis and the line y = 2
    serve as boundaries of the
    feasible region. They intersect at
    (0, 2).

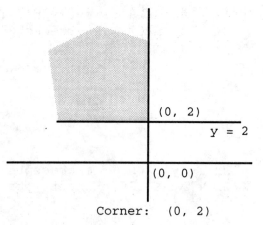

Corner: (0, 2)

15. $-x + 2y = 6$ Intercepts: (-6, 0) and (0, 3).

    $3x + 2y = -2$ Intercepts: $(-\frac{2}{3}, 0)$ and (0, -1).

    The lines intersect at (-2, 2).

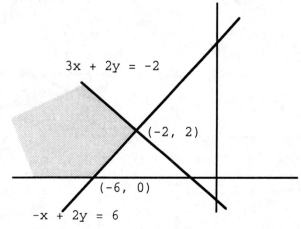

Corners: (-2, 2) and (-6, 0)

17. $3x + 4y > 24$ Intercepts: (8,0),(0,6)
    $4x + 5y < 20$ Intercepts: (5,0),(0,4)
    $x > 0$

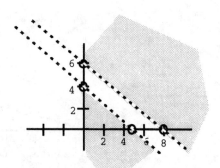

No feasible solution

19. $x + y \le 2$ Intercepts: $(2,0),(0,2)$
        $2x + y \ge 6$ Intercepts: $(3,0),(0,6)$
        $y \ge 0$

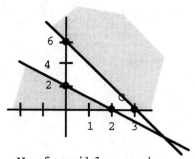

No feasible region

21. $x + y \ge 1$ Intercepts: $(1,0),(0,1)$
        $-x + y \ge 2$ Intercepts: $(-2,0),(0,2)$
        $5x - y \le 4$ Intercepts: $(4/5,0),(0,-4)$

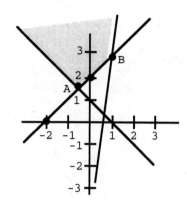

Corners: $A(-1/2,3/2)$
         $B(3/2,7/2)$

23.   Graph the lines as shown. The lines $-x + y = 5$ and $6x + 6y = 48$
       intersect at $(1, 6)$. The lines $-x + y = 5$ and $5x + 6y = -14$
       intersect at $(-4, 1)$. The lines $5x + 6y = -14$ and $-4x + 6y = -32$
       intersect at $(2, -4)$. The lines $-4x + 6y = -32$ and $6x + 7y = 48$
       intersect at $(8, 0)$.

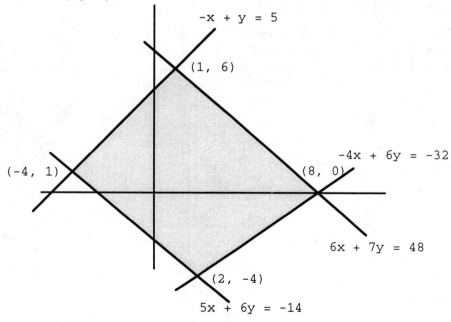

Corners: $(-4, 1)$, $(1, 6)$,
          $(8, 0)$, and $(2, -4)$

25. $3x + 4y \leq 12$ Intercepts: $(4, 0)$, $(0, 3)$
     $5x + 3y \leq 15$ Intercepts: $(3, 0)$, $(0, 5)$

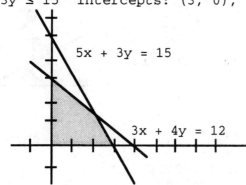

     $5x + 3y = 15$

     $3x + 4y = 12$

Bounded feasible region

27. $x + y \geq 6$ Intercepts: $(6, 0)$ $(0, 6)$
     $3x + 6y \geq 24$ Intercepts: $(8, 0)$ $(0, 4)$
     $x \geq 0, y \geq 0$

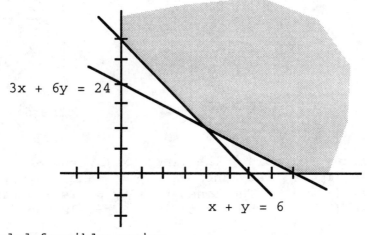

     $3x + 6y = 24$

     $x + y = 6$

Unbounded feasible region

29. $-2x + y \leq 2$ Intercepts: $(-1,0)$, $(0,2)$
      $3x + y \leq 3$ Intercepts: $(1,0)$, $(0,3)$
      $-x + y \geq -4$ Intercepts: $(4,0)$, $(0,-4)$
       $x + y \geq -3$ Intercepts: $(-3,0)$, $(0,-3)$

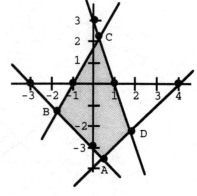

Corners: A$(1/2,-7/2)$
         B$(-5/3,-4/3)$
         C$(1/5,12/5)$
         D$(7/4,-9/4)$

31. Let x = amount of High Fibre, y = amount of Corn Bits.
            $.25x + .02y \geq .40$
            $.04x + .10y \geq .25$
              $x \geq 0, y \geq 0$

33.   Let x = number of student tickets, y = number of adult tickets.
$$x + y \geq 500$$
$$3x + 8y \geq 2500$$
$$x \geq 0, y \geq 0$$

35.   Let x = number correct, y = number incorrect.
$$x + y \geq 60$$
$$4x - y \geq 200$$
$$x \geq 0, y \geq 0$$

37.   Let x = number of balcony tickets, y = number of main floor tickets
Total tickets $x + y \geq 3000$
Main floor $y \geq 1200$
Sales $15x + 25y \geq 60,000$

39. $7x + 3y \leq 21$ Intercepts: (3, 0) (0, 7)
$5x + 4y \leq 20$ Intercepts: (4, 0) (0, 5)

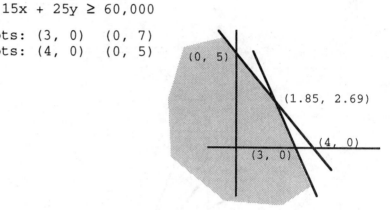

Corner (1.85, 2.69)

41. $9x + 4y \leq 36$
$6x + 5y \leq 30$
$5x + 10y \leq 50$

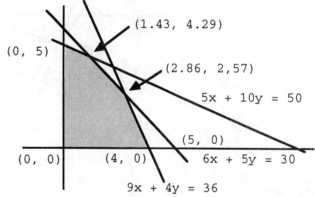

Corners: (0,0) (0, 5) (1.43, 4.29) (2.86, 2.57) (4, 0)

43. $9.5x + 3.2y \geq 30.4$
       $6.75x + 7.5y \geq 50.625$
       $x \geq 0, y \geq 0$

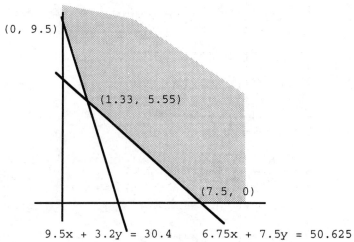

$9.5x + 3.2y = 30.4$ $6.75x + 7.5y = 50.625$

Corners: (0, 9.5) (1.33, 5.55) (7.5, 0)

45. $2.86x + 1.19y \leq 14.29$
       $1.94x + 4.17y \leq 22.83$
       $3.33x + 2.50y \leq 18.00$
       $x \geq 0, y \geq 0$

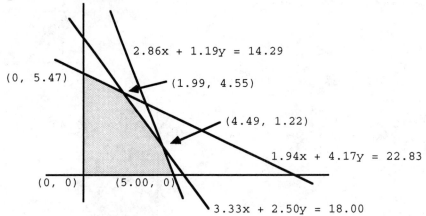

Corners: (0, 0) (0, 5.47) (1.99, 4.55) (4.49, 1.22) (5.00, 0)

Section 3.3

1.   Let x = number of style A, y = number of style B.
      Maximize z = 50x + 40y Subject to: $x + y \leq 80$
                                             $x + 2y \leq 110$
                                             $x \geq 0, y \geq 0$

3. $3x + 2y \leq 18$ Intercepts: (6, 0),(0, 9)
      $3x + y \leq 15$ Intercepts: (5, 0),(0, 15)

	z = 20x + 12y
A(0, 0)	0
B(0, 9)	108
C(4, 3)	116
D(5, 0)	100

Maximum is 116 at (4, 3)

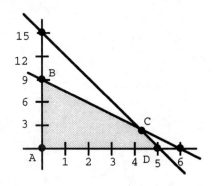

5.   5x + y ≤ 35   Intercepts: (7, 0),(0, 35)
      3x + y ≤ 27   Intercepts: (9, 0),(0, 27)
      _____|___z_=_9x_+_2y_____
      A(0, 0)| 0
      B(0, 27)| 54
      C(4, 15)| 66
      D(7, 0)| 63
      Maximum is 66 at (4, 15)

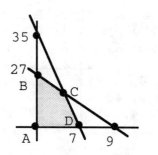

7.   4x + y ≥ 40   Intercepts: (10,0), (0,40)
      4x + 3y ≥ 64   Intercepts: (16,0), (0,64/3)
      _____|___z_=_2x_+_3y_____
      A(0,40)| 120
      B(7,12)| 50
      C(16,0)| 32
      Minimum is 32 at (16, 0)

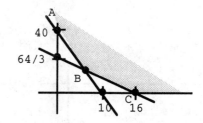

9.   10x + 11y ≤ 330   Intercepts: (33, 0),(0, 30)
       4x + 6y ≤ 156   Intercepts: (39, 0),(0, 26)
      _____|___z_=_9x_+_13y_____
      A(0, 0)| 0
      B(0, 26)| 338
      C(33/2, 15)| 343.5
      D(33, 0)| 297
      Maximum is 343.5 at (16.5, 15)

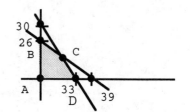

11.   -x + 2y ≤ 40   Intercepts: (-40, 0),(0, 20)
       x + 4y ≤ 54   Intercepts: (54, 0),(0, 27/2)
      3x + y ≤ 63   Intercepts: (21, 0),(0, 63)
      _____|___z_=_20x_+_30y_____
      A(0, 0)| 0
      B(0, 27/2)| 405
      C(18, 9)| 630
      D(21, 0)| 420
      Maximum is 630 at (18, 9)

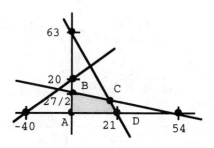

13.   x + 5y ≤ 250   Intercepts: (250, 0),(0, 50)
      2x + 5y ≤ 300   Intercepts: (150, 0),(0, 60)
       x ≤ 75
      _____|___z_=_320x_+_140y_____
      A(0, 0)| 0
      B(0, 50)| 7,000
      C(50, 40)| 21,600
      D(75, 30)| 28,200
      E(75, 0)| 24,000
      Maximum is 28,200 at (75, 30)

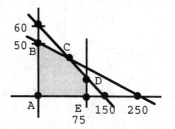

15. 6x + y ≥ 52   Intercepts: (26/3, 0),(0, 52)
     2x + y ≥ 20   Intercepts: (10, 0),(0, 20)
      x + 4y ≥ 24   Intercepts: (24, 0),(0, 6)

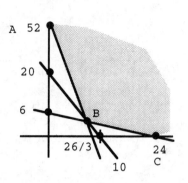

	z = 5x + 3y
A(0, 52)	156
B(8, 4)	52
C(24, 0)	120

Minimum is 52 at (8, 4)

17. 3x + y ≥ 30   Intercepts: (10, 0),(0, 30)
     4x + 3y ≥ 60   Intercepts: (15, 0),(0, 20)
      x + 2y ≥ 20   Intercepts: (20, 0),(0, 10)

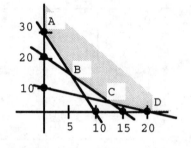

	z = 5x + 3y
A(0, 30)	90
B(6, 12)	66
C(12, 4)	72
D(20, 0)	100

Minimum is 66 at (6, 12)

19. 2x + 10y ≤ 80   Intercepts: (40, 0),(0, 8)
     6x + 2y ≤ 72   Intercepts: (12, 0),(0, 36)
     3x + 2y ≥ 6   Intercepts: (2, 0),(0, 3)

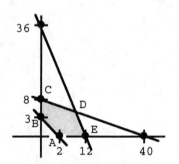

	z = 20x + 30y
A(2, 0)	40
B(0, 3)	90
C(0, 8)	240
D(10, 6)	380
E(12, 0)	240

Maximum is 380 at (10, 6)
Minimum is 40 at (2, 0)

21.   6x + 8y ≤ 300   Intercepts: (50, 0), (0, 37.5)
     15x + 22y ≥ 330   Intercepts: (22, 0), (0, 15)
      x ≤ 40, y ≤ 21

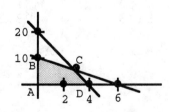

	z = 5x + 6y
A(0, 15)	90
B(0, 21)	126
C(22, 21)	236
D(40, 7.5)	245
E(40, 0)	200
F(22 ,0)	110

Maximum is 245 at (40, 7.5)
Minimum is 90 at (0, 15)

23. 5x + 3y ≤ 30   Intercepts: (6, 0), (0, 10)
     5x + y ≤ 20   Intercepts: (4, 0), (0, 20)

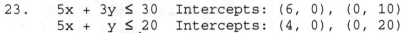

	z = 15x + 9y
A(0, 0)	0
B(0, 10)	90
C(3, 5)	90
D(4, 0)	60

Maximum is 90 at any point on line
5x + 3y = 30 from (0, 10) to (3, 5)

25. $3x + 2y \geq 60$ Intercepts: (20, 0),(0, 30)
       $10x + 3y \leq 180$ Intercepts: (18, 0),(0, 60)
       $y \leq 24$

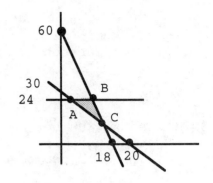

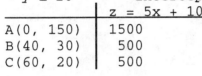

	$z = 9x + 6y$
A(4, 24)	180
B(10.8, 24)	241.2
C($\frac{180}{11}$, $\frac{60}{11}$)	180

Minimum is 180 at (4, 24),
($\frac{180}{11}$, $\frac{60}{11}$) and points between

27. $3x + y \geq 150$ Intercepts: (50, 0)(0, 150)
        $x + 2y \geq 100$ Intercepts: (100, 0)(0, 50)
         $y \geq 20$ Intercepts: (0, 20)

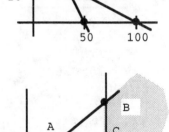

	$z = 5x + 10y$
A(0, 150)	1500
B(40, 30)	500
C(60, 20)	500

Minimum is 500 at (40, 30), (60, 20) and
points between

29. $2x - 3y \geq 10$ Intercepts: (5,0),(0, -10/3)
       $x \geq 8$

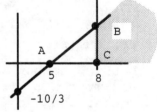

Unbounded feasible region, no maximum

31. $-3x + y \leq 4$ Intercepts: (-4/3, 0),(0, 4)
       $-2x - y \geq 1$ Intercepts: (-1/2, 0),(0, -1)

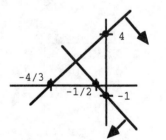

No feasible region, no minimum

33. $3x - y \leq 8$ Intercepts: (8/3, 0),(0, -8)
       $x - 2y \geq 5$ Intercepts: (5, 0),(0, -5/2)

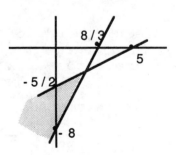

No feasible region

35.   Let x = number of standard VCR's, y = number of deluxe VCR's.
          Maximize z = 39x + 26y
          Subject to $8x + 9y \leq 2200$
                       $115x + 136y \leq 18,000$
                       $x \geq 35, y \geq 0$

59

37. Let x = number shipped to A, y = number shipped to B.

    Minimize C = 13x + 11y

    Subject to x + y ≥ 250

    x + 30 ≤ 140

    y + 18 ≤ 165

    x ≥ 0, y ≥ 0

39. Let x = number cartons of regular, y = number cartons of diet.

    Maximize z = .15x + .17y

    Subject to x + 1.2y ≤ 5400

    x +   y ≤ 5000

    x ≥ 0, y ≥ 0

41. Let x = number of desk lamps y = number of floor lamps.

    Maximize z = 2.65x + 3.15y   Intercepts

    Subject to .8x + y ≤ 1200   (1500, 0) (0, 1200)

    4x + 3y ≤ 4200   (1050, 0) (0, 1400)

    x ≥ 0, y ≥ 0

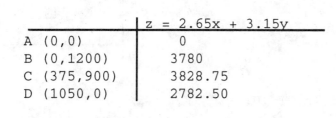

	z = 2.65x + 3.15y
A (0,0)	0
B (0,1200)	3780
C (375,900)	3828.75
D (1050,0)	2782.50

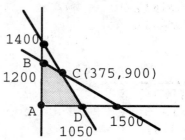

Maximum profit is $3828.75 at 375 desk lamps, 900 floor lamps.

43. Let x = number acres for cattle, y = number acres for sheep.

    Maximize z = 30x + 32y

    Subject to   x + y ≤ 1000 Intercepts: (1000, 0),(0, 1000)

    2x + 8y ≤ 3200 Intercepts: (1600, 0),(0, 400)

    x ≥ 0, y ≥ 0

	z = 30x +32y
A(0, 0)	0
B(0, 400)	12,800
C(800, 200)	30,400
D(1000, 0)	30,000

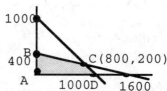

Maximum profit is $30,400 for 800 acres for cattle and 200 acres for sheep.

45. Let x = weight of food I, y = weight of food II.

    Minimize z = .03x + .04y

    Subject to   .4x + .6y ≥ 10   Intercepts: (25, 0),(0, 50/3)

    .5x + .2y ≥ 7.5   Intercepts: (15, 0),(0, 37.5)

    .06x + .04y ≥ 1.2   Intercepts: (20, 0),(0, 30)

    x ≥ 0, y ≥ 0

	z = .03x + .04y
A (0, 37.5)	1.50
B (7.5, 18.75)	.975
C (16, 6)	.72
D (25, 0)	.75

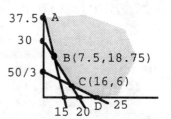

Minimum cost is $.72 using 16 g of food I and 6 g of food II.

60

47. Let x = number of standard gears, y = number of heavy duty gears.
    Minimize z = 15x + 22y
    subject to   8x + 10y ≥ 12,000   Intercepts: (1500, 0),(0, 1200)
                 3x + 10y ≥ 8,400   Intercepts: (2800, 0),(0, 840)
                  x ≥ 0, y ≥ 0

	z = 15x + 22y
A (0,1200)	26,400
B (720,624)	24,528
C (2800,0)	42,000

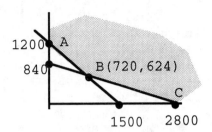

Minimum cost is $24,528 producing 720 standard gears and 624 heavy duty gears.

49. Let x = number of SE vans, y = number of LE vans.
    Minimize z = 2700x + 2400y
    subject to   x + y ≥ 9   Intercepts: (9, 0),(0, 9)
              16,000x + 20,000y ≤ 160,000   Intercepts: (10, 0),(0, 8)
                  x ≥ 0, y ≥ 0

	2700x + 2400y
A(9,0)	24,300
B(10,0)	27,000
C(5,4)	23,100

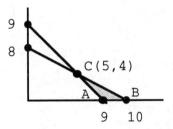

Minimum maintenance cost = $23,100 with 5 SE vans and 4 LE vans.

51. (a) Let x = cartons of regular, y = cartons of diet.
         Maximize z = .15x + .17y
         Subject to   x + 1.2y ≤ 5400
                      x + y ≤ 5000
                      x ≥ 0, y ≥ 0

Corners	z = .15x + .17y
A (0,0)	0
B (0,4500)	765
C (3000,20000)	790
D (5000,0)	750

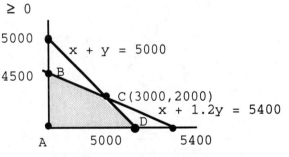

Maximum profit is $790 with 3,000 cartons of regular and 2,000 cartons of diet.

61

(b)   (i)   The objective function becomes z = .20x + .17y, the constraints and corners remain the same.

Corners	z = .20x + .17y
A(0,0)	0
B(0,4500)	765
C(3000,2000)	940
D(5000,0)	1000

The optimal solution changes to the corner (5000, 0) and the maximum profit increases to $1000.

(ii) The objective function becomes z = .15x + .19y, the constraints and corners remain the same

Corners	.15x + .19y
A(0,0)	0
B(0,4500)	855
C(3000,2000)	830
D(5000,0)	750

The optimal solution changes to the corner (0, 4500) and the maximum profit increases to $855.

(iii) The objective function becomes z = .15x + .16y, the constraints and corners remain the same

Corners	.15x + .16y
A(0,0)	0
B(0,4500)	720
C(3000,2000)	770
D(5000,0)	750

The optimal solution remains at the corner (3000,2000) but the maximum profit decreases to $770.

53.   Let x = number of square feet of type A, y = number of square feet of type B glass.

(a)   Minimize z = x + .25y subject to

$$x + \quad y \geq 4000 \quad \text{Intercepts: } (4000, 0),(0, 4000)$$
$$.8x + 1.2y \leq 4500 \quad \text{Intercepts: } (5625, 0),(0, 3750)$$
$$x \geq 0, \, y \geq 0$$

	z = x + .25y
A(4000, 0)	4000
B(750, 3250)	1562.5
C(5625, 0)	5625

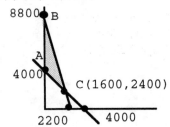

Minimum conductance = 1562.5 using 750 square feet of type A glass and 3250 square feet of type B.

(b)   Minimize z = .8x + 1.2y, subject to

x + .25y ≤ 220 Intercepts: (2200, 0),(0, 8800)
x +   y ≥ 400 Intercepts: (4000, 0),(0, 4000)

	z = .8x + 1.2y
A(0, 4000)	4,800
B(0, 8800)	10,560
C(1600, 2400)	4,160

Minimum cost is $4,160 using 1600 sq. ft. of type A and 2400 sq. ft. of type B.

55. Let x = number of days the Glen Echo plant operates,
    y = number of days the Speegleville Road plant operates.
    Minimize z = x + y, subject to
$$200x + 300y \geq 2400 \quad \text{Intercepts: } (12, 0), (0, 8)$$
$$300x + 200y \geq 2100 \quad \text{Intercepts: } (7, 0), (0, 10.5)$$
$$x \geq 0, \ y \geq 0$$

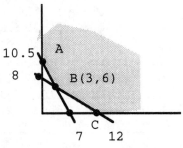

	z = x + y
A(0, 10.5)	10.5
B(3, 6)	9
C(12, 0)	12

Minimum number of days is 9
with 3 at Glen Echo and 6 at
Speegleville Road.

57. Let x = number of computers from Supplier A to Raleigh.
    Let y = number of computers from Supplier A to Greensboro.
    Then 165 − x = number of computers from Supplier B to Raleigh.
        190 − y = number of computers from Supplier B to Greensboro.
    We summarize this information in the following table.

		Raleigh	Greensboro	
A	cost (ea.)	$35	$30	200
	Number	x	y	available
B	cost (ea.)	$45	$50	230
	Number	165−x	190−y	available
		165 needed	190 needed	

The shipping cost total
$$C = 35x + 30y + 45(165 - x) + 50(190 - y)$$
$$= -10x - 20y + 16925$$
The constraints are
$$x + y \leq 200 \quad \text{(available from A)}$$
$$165 - x + 190 - y \leq 230 \quad \text{(available from B) which}$$
reduces to 125 ≤ x + y
$$x \geq 0, \ y \geq 0, \ x \leq 165, \ y \leq 190$$
The feasible region and corner points for this problem are the
following
Compute C = 16925 − 10x − 20y at each corner.

Corner	C
(0, 125)	14425
(0, 190)	13125
(10, 190)	13025
(165, 35)	14575
(165, 0)	15275
(125, 0)	15675

The minimum cost is $13,025 and occurs when x = 10 and y = 190.
Thus, Supplier A should ship 10 to Raleigh and 190 to Greensboro.
Supplier B should ship 155 to Raleigh and none to Greensboro.

59. Let x = number of lamps shipped from A to Emporia.
     Let y = number of lamps shipped from A to Ardmore.
     Then
          25 - x = number of lamps shipped from B to Emporia.
          20 - y = number of lamps shipped from B to Ardmore.
     We use the following table to summarize this information.

		To		
		Emporia	Ardmore	
A	cost (ea.)	$5	$6	10
	Number	x	y	available
B	cost (ea.)	$7	$4	40
	Number	25 - x	20 - y	available
		25 needed	20 needed	

The total shipping costs are
$$C = 5x + 6y + 7(25 - x) + 4(20 - y)$$
$$= -2x + 2y + 255$$
The constraints are
     $x + y \le 10$ (available from A), and
$25 - x + 20 - y \le 40$ (available from B), which reduces to
     $5 \le x + y$.
     $x \ge 0, y \ge 0, x \le 25, y \le 20$.
Here is the feasible region and the corners:
Calculate C = 255 - 2x + 2y at the corners.

Corner	C
(0, 5)	265
(0, 10)	275
(10, 0)	235
(5, 0)	245

The minimum cost is $235 and it occurs when x = 10 and y = 0.
Thus, Supplier A should ship 10 to Emporia and none to Ardmore.
Supplier B should ship 15 to Emporia and 20 to Ardmore.

61. In a maximization problem the profit line is moved away from the
     origin until a corner point is the only point it touches in the
     feasible region. To find the optimal integer solution move the
     profit line outward until it touches the last point in the feasible
     region with integer coordinates. In this case that is the point
     (6, 1) so the optimal profit occurs when x = 6 and y = 1.

63. The corners of the feasible region and the corresponding values of
     z are

Corner	z = 5.2x + 7.3y
(0, 8.32)	60.74
(3.87, 0)	20.12
(1.99, 6.03)	54.37

The maximum z is 60.74 and occurs at (0, 8.32).

65. The corners of the feasible region and the corresponding values of z are

Corner	$z = 2.2x + 4.8y$
(0, 0)	0
(0, 6.6)	31.68
(3, 0)	6.6

The maximum value of z is 31.68 and occurs at (0, 6.6).

3.4 Exercises

1. Let x_1 = number of acres of onions
   x_2 = number of acres of carrots
   x_3 = number of acres of lettuce

   The linear programming problem is then stated in the following way:
   Maximize profit
   $$z = 65x_1 + 70x_2 + 50x_3 \text{ subject to}$$
   $$x_1 + x_2 + x_3 \le 75 \text{ (acreage available)}$$
   $$250x_1 + 300x_2 + 325x_3 \le 225,000 \text{ (production costs)}$$
   $$x_1 \ge 0, x_2 \ge 0, x_3 \ge 0$$

3. The linear programming problem is the following:
   Let x_1 = number of chests, x_2 = number of desks, and
   x_3 = number of silverware boxes.
   Maximize profit
   $$z = 180x_1 + 300x_2 + 45x_3 \text{ subject to}$$
   $$270x_1 + 310x_2 + 90x_3 \le 5,000 \text{ (cost limitation)}$$
   $$7x_1 + 18x_2 + 1.5x_3 \le 1500 \text{ (Space limitations)}$$
   $$x_1 + x_2 + x_3 \le 200 \text{ (Maximum number allowed)}$$
   $$x_1 \ge 0, x_2 \ge 0, x_3 \ge 0$$

5. Let x_1 = number pounds of Lite
   x_2 = number pounds of Trim
   x_3 = number pounds of Regular
   x_4 = number pounds of Health Fare
   Maximize profit
   $$z = .25x_1 + .25x_2 + .27x_3 + .32x_4 \text{ subject to}$$
   $$.75x_1 + .50x_2 + .80x_3 + .15x_4 \le 2400 \text{ (Wheat)}$$
   $$.20x_1 + .25x_2 + .20x_3 + .50x_4 \le 1400 \text{ (Oats)}$$
   $$.05x_1 + .20x_2 + .25x_4 \le 700 \text{ (Raisins)}$$
   $$.05x_2 + .10x_4 \le 250 \text{ (Nuts)}$$
   $$x_1 \ge 0, x_2 \ge 0, x_2 \ge 0, x_4 \ge 0$$

7. The following figure shows the possible two-day and three-day periods with the periods labeled x_1, x_2, ..., x_5, which also represent the number of computers needed for that time period.

	2-day rental		3-day rental	
Day 1	x_1			
Day 2		x_2	x_4	
Day 3			x_3	x_5
Day 4				

We can state this as the following linear programming problem.
Minimize rent
$$z = 80x_1 + 80x_2 + 80x_3 + 100x_4 + 100x_5 \quad \text{subject to}$$
$$x_1 + \qquad\qquad x_4 \geq 12 \quad (\text{\# needed first day})$$
$$x_1 + x_2 + \qquad x_4 + x_5 \geq 15 \quad (\text{\# needed second day})$$
$$x_2 + x_3 + x_4 + x_5 \geq 18 \quad (\text{\# needed third day})$$
$$x_3 + \qquad x_5 \geq 20 \quad (\text{\# needed fourth day})$$
$$x_1 \geq 0,\ x_2 \geq 0,\ x_3 \geq 0,\ x_4 \geq 0,\ x_5 \geq 0$$

9. The following chart summarizes the given information.

	Plantation			
	P-1	P-2	P-3	
Plant A	Cost = 50 No. = x_1	Cost = 65 No. = x_2	Cost = 58 No. = x_3	Process at least 540
Plant B	Cost = 40 No. = x_4	Cost = 55 No. = x_5	Cost = 69 No. = x_6	Process at least 450
	Produces 250	Produces 275	Produces 310	

The linear programming problem is the following.
Minimize shipping costs
$$z = 50x_1 + 65x_2 + 58x_3 + 40x_4 + 55x_5 + 69x_6 \quad \text{subject to}$$
$$x_1 + x_2 + x_3 \geq 540 \quad (\text{Plant A capacity})$$
$$x_4 + x_5 + x_6 \geq 450 \quad (\text{Plant B capacity})$$
$$x_1 + x_4 \geq 250 \quad (\text{P-1 production})$$
$$x_2 + x_5 \geq 275 \quad (\text{P-2 production})$$
$$x_3 + x_6 \geq 310 \quad (\text{P-3 production})$$
$$x_1 \geq 0,\ x_2 \geq 0,\ x_3 \geq 0,\ x_4 \geq 0,\ x_5 \geq 0,\ x_6 > 0.$$

11. Let x_1 = number of days Red Mountain operates
x_2 = number of days Cahaba operates
x_3 = number of days Clear Creek operates
The objective function is the cost function.
Minimize $z = 8000x_1 + 14000x_2 + 12000x_3$ subject to
$$105x_1 + 295x_2 + 270x_3 \geq 1800 \quad (\text{amount of low-grade})$$
$$90x_1 + 200x_2 + 85x_3 \geq 1350 \quad (\text{amount of high-grade})$$
$$x_1 \geq 0,\ x_2 \geq 0,\ x_3 \geq 0$$

13.   Let x_1 = number of Mini Packets
            x_2 = number of Mid Packets
            x_3 = number of Maxi Packets
       Minimize costs
            $z = 19x_1 + 30x_2 + 45x_3$ subject to
            $8x_1 + 7x_2 + 6x_3 \geq 260$ (Number of "A" tickets)
            $2x_1 + 7x_2 + 14x_3 \geq 175$ (Number of "B" tickets)
            $x_1 \geq 0, x_2 \geq 0, x_3 \geq 0$

15.   Let x_1 = number of servings of milk
            x_2 = number of servings of vegetables
            x_3 = number of servings of fruit
            x_4 = number of servings of bread
            x_5 = number of servings of meat
       Maximize carbohydrates
            $z = 12x_1 + 7x_2 + 11x_3 + 15x_4$ subject to
            $8x_1 + 2x_2 + 2x_4 + 7x_5 \leq 35$ (amount of protein)
            $10x_1 + x_4 + 5x_5 \leq 40$ (amount of fat)
            $x_1 \geq 0, x_2 \geq 0, x_3 \geq 0, x_4 \geq 0, x_5 \geq 0$

17.   Let x_1 = percent invested in A bonds
            x_2 = percent invested in AA bonds
            x_3 = percent invested in AAA bonds
       Maximize return
            $z = .072x_1 + .068x_2 + .065x_3$ subject to
            $x_1 + x_2 + x_3 \leq 100$ (Investment Total)
            $x_1 + x_2 \leq 65$ (Invested in A and AA)
            $x_2 + x_3 \geq 50$ (Invested in AA and AAA)
            $x_1 \geq 0, x_2 \geq 0, x_3 \geq 0$

19.   Let x_1 = number of sheets of cutting plan 1, x_2 = number of sheets
       of cutting plan 2, etc. The waste to be minimized is
       waste = $x_2 + 3x_3 + 2x_4 + 12x_5 + 14x_7$ and the constraints are
            $3x_2 + 9x_3 + 6x_4 + 6x_6 \geq 165$ (no. of 15 inch doors)
            $9x_1 + 6x_2 + 3x_4 + 3x_7 \geq 200$ (no. of 16 inch doors)
            $6x_5 + 3x_6 + 3x_7 \geq 85$ (no. of 18 inch doors)
            $x_1 \geq 0, x_2 \geq 0, x_3 \geq 0, x_4 \geq 0, x_5 \geq 0, x_6 \geq 0$

Review Exercises,   Chapter 3

1.   (a) $5x + 7y < 70$ (b) $2x - 3y > 18$
       Intercepts: (14, 0), (0, 10)   Intercepts: (9, 0), (0, -6)

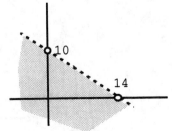
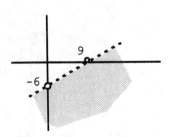

(c)   x + 9y ≤ 21
      Intercepts:
      (21, 0), (0, 7/3)

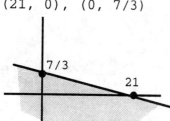

(d)   -2x + 12y ≥ 26
      Intercepts:
      (-13, 0), (0, 13/6)

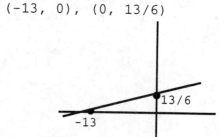

(e)   y ≥ -6

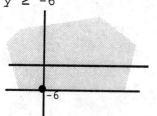

(f)   x ≤ 3

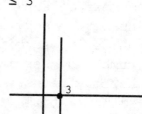

3.   x - 3y ≥ 6   Intercepts: (6, 0),(0, -2)
      x - y ≤ 4   Intercepts: (4, 0),(0, -4)
           y ≥ -5   Intercepts: (0, -5)
           Corners: A(-1, -5)
                         B(-9, -5)
                         C(3, -1)

5.   -3x + 4y ≤ 20   Intercepts: (-20/3,0),(0,5)
        x + y ≥ -2   Intercepts: (-2, 0),(0, -2)
       8x + y ≤ 40   Intercepts: (5, 0),(0, 40)
            Corners: A(-4, 2)
                          B(4, 8)
                          C(5, 0)
                          D(-2, 0)

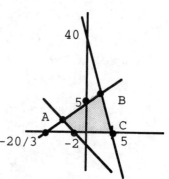

7.   x - 2y ≤ 0   Intercepts: (0, 0),(2, 1)
     -2x + y ≤ 2   Intercepts: (-1, 0),(0, 2)
         x ≤ 2   Intercepts: (2, 0)
         y ≤ 2   Intercepts: (0, 2)
            Corners: A(-4/3, -2/3)
                        B(0, 2)
                        C(2, 2)
                        D(2, 1)

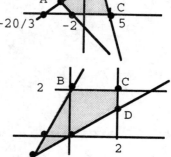

9.   3x + 2y ≤ 12   Intercepts: (4, 0),(0, 6)
       x + y ≤ 5   Intercepts: (5, 0),(0, 5)

	5x + 4y
A(0, 0)	0
B(0, 5)	20
C(2, 3)	22
D(4, 0)	20

Maximum z is 22 at (2, 3)

11. $3x + 2y \geq 18$ Intercepts: (6, 0),(0, 9)
        $x + 2y \geq 10$ Intercepts: (10, 0),(0, 5)
       $5x + 6y \geq 46$ Intercepts: (9.2, 0),(0, 23/3)

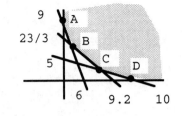

	$5x + 4y$	$10x + 12y$
A(0, 9)	36	108
B(2, 6)	34	92
C(8, 1)	44	92
D(10, 0)	50	100

   (a)   Min is 34 at (2, 6)
   (b)   Min is 92, at any point on line $5x + 6y = 46$
          from (2, 6) to (8, 1)

13. $2x + y \leq 90$ Intercepts: (45, 0),(0, 90)
       $x + 2y \leq 80$ Intercepts: (80, 0),(0, 40)
       $x + y \leq 50$ Intercepts: (50, 0),(0, 50)

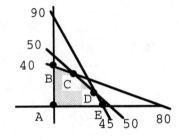

	$4x + 7y$
A(0, 0)	0
B(0, 40)	280
C(20, 30)	290
D(40, 10)	230
E(45, 0)	180

Maximum z is 290 at (20, 30)

15.   Let x = number of hours Line A is used, y = number of hours Line B
       is used.
       (a) $65x + 105y \leq 1500(8)$ (b)
              (For an 8-hour day)
              $x \geq 0,\ y \geq 0$

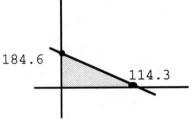

17.   Let x = number of adult tickets, y = number of children tickets.
                    $x + y \leq 275$
                $4.5x + 3y \geq 1100$
                    $x \geq 0,\ y \geq 0$

19.   Maximize $r = 90x + 100y$ subject to
           $(.9)(100)x + (.8)(100)y \leq 80,000$
                   Intercepts: (888.9, 0),(0, 1000)
           $(.1)(100)x + (.2)(100)y \leq 12,000$
                   Intercepts: (1200, 0),(0, 600)
              $x \geq 0,\ y \geq 0$

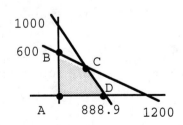

	$90x + 100y$
A(0, 0)	0
B(0, 600)	60,000
C(640, 280)	85,600
D(888.9, 0)	80,000

Maximum revenue is $85,600 when 640 bars of standard and 280 of
premium are produced.

69

21. Let x_1 = number of A-teams
     x_2 = number of B-teams
     x_3 = number of C-teams
     Maximize number of inoculations
     $z = 175x_1 + 110x_2 + 85x_3$ subject to
     $x_1 + x_2 + x_3 \leq 75$ (Number of doctors)
     $3x_1 + 2x_2 + x_3 \leq 200$ (Number of nurses)
     $x_1 \geq 0,\ x_2 \geq 0,\ x_3 \geq 0$

23. Let x_1 = number of Type I pattern
     x_2 = number of Type II pattern
     x_3 = number of Type III pattern
     x_4 = number of Type IV pattern
     Maximize profit
     $z = 48x_1 + 45x_2 + 55x_3 + 65x_4$ subject to
     $40x_1 + 25x_2 + 30x_3 + 45x_4 \leq 1250$ (Number tulips)
     $25x_1 + 50x_2 + 40x_3 + 45x_4 \leq 1600$ (Number daffodils)
     $6x_1 + 4x_2 + 8x_3 + 2x_4 \leq 195$ (Number boxwood)
     $x_1 \geq 0,\ x_2 \geq 0,\ x_3 \geq 0,\ x_4 \geq 0$

Chapter 4
Linear Programming: The Simplex Method

Section 4.1

1. $2x_1 + 3x_2 + s_1 = 9$
   $x_1 + 5x_2 + s_2 = 16$

3. $x_1 + 7x_2 - 4x_3 + s_1 = 150$
   $5x_1 + 9x_2 + 2x_3 + s_2 = 435$
   $8x_1 - 3x_2 + 16x_3 + s_3 = 345$

5. $2x_1 + 6x_2 + s_1 = 9$
   $x_1 - 5x_2 + s_2 = 14$
   $-3x_1 + x_2 + s_3 = 8$
   $-3x_1 - 7x_2 + z = 0$

7. $6x_1 + 7x_2 + 12x_3 + s_1 = 50$
   $4x_1 + 18x_2 + 9x_3 + s_2 = 85$
   $x_1 - 2x_2 + 14x_3 + s_3 = 66$
   $-420x_1 - 260x_2 - 50x_3 + z = 0$

9. $$\left[\begin{array}{ccccc|c} 4 & 5 & 1 & 0 & 0 & 10 \\ 3 & 1 & 0 & 1 & 0 & 25 \\ \hline -3 & -17 & 0 & 0 & 1 & 0 \end{array}\right]$$

11. $$\left[\begin{array}{ccccccc|c} 16 & -4 & 9 & 1 & 0 & 0 & 0 & 128 \\ 8 & 13 & 22 & 0 & 1 & 0 & 0 & 144 \\ 5 & 6 & -15 & 0 & 0 & 1 & 0 & 225 \\ \hline -20 & -45 & -40 & 0 & 0 & 0 & 1 & 0 \end{array}\right]$$

13. Let x_1 = number of cartons of screwdrivers, x_2 = number of cartons of chisels, x_3 = number of cartons of putty knives.
    Maximize $z = 5x_1 + 6x_2 + 5x_3$
    Subject to $3x_1 + 4x_2 + 5x_3 \leq 2200$
    $15x_1 + 12x_2 + 11x_3 \leq 8500$
    $x_1 \geq 0, x_2 \geq 0, x_3 \geq 0$
    $$\left[\begin{array}{cccccc|c} 3 & 4 & 5 & 1 & 0 & 0 & 2200 \\ 15 & 12 & 11 & 0 & 1 & 0 & 8500 \\ \hline -5 & -6 & -5 & 0 & 0 & 1 & 0 \end{array}\right]$$

15. Let x_1 = amount of salad, x_2 = amount of potatoes, x_3 = amount of steak.
    Maximize $z = 0.5x_1 + 1x_2 + 9x_3$
    Subject to $20x_1 + 50x_2 + 56x_3 \leq 1000$
    $1.5x_1 + 3x_2 + 2x_3 \leq 35$
    $x_1 \geq 0, x_2 \geq 0, x_3 \geq 0$
    $$\left[\begin{array}{cccccc|c} 20 & 50 & 56 & 1 & 0 & 0 & 1000 \\ 1.5 & 3 & 2 & 0 & 1 & 0 & 35 \\ \hline -0.5 & -1 & -9 & 0 & 0 & 1 & 0 \end{array}\right]$$

17. Let x_1 = lbs of Lite, x_2 = lbs. of Trim, x_3 = lbs. of Health Fare.
    Maximize $z = .25x_1 + .25x_2 + .32x_3$
    Subject to $.75x_1 + .50x_2 + .15x_3 \leq 2320$ (Wheat)
    $.25x_1 + .25x_2 + .60x_3 \leq 1380$ (Oats)
    $.25x_2 + .25x_3 \leq 700$ (Raisins)
    $x_1 \geq 0, x_2 \geq 0, x_3 \geq 0$

71

$$\begin{bmatrix} .75 & .50 & .15 & 1 & 0 & 0 & 0 & 2320 \\ .25 & .25 & .60 & 0 & 1 & 0 & 0 & 1380 \\ 0 & .25 & .25 & 0 & 0 & 1 & 0 & 700 \\ -.25 & -.25 & -.32 & 0 & 0 & 0 & 1 & 0 \end{bmatrix}$$

19. Let x_1 = number of military trunks, x_2 = number of commercial trunks, x_3 = number of decorative trunks.

Maximize $z = 6x_1 + 7x_2 + 9x_3$

Subject to
$$4x_1 + 3x_2 + 2x_3 \le 4900 \quad \text{(Assembly)}$$
$$x_1 + 2x_2 + 4x_3 \le 2200 \quad \text{(Finishing)}$$
$$.1x_1 + .2x_2 + .3x_3 \le 210 \quad \text{(Packing)}$$
$$x_1 \ge 0, \ x_2 \ge 0, \ x_3 \ge 0$$

$$\begin{bmatrix} 4 & 3 & 2 & 1 & 0 & 0 & 0 & 4900 \\ 1 & 2 & 4 & 0 & 1 & 0 & 0 & 2200 \\ .1 & .2 & .3 & 0 & 0 & 1 & 0 & 210 \\ -6 & -7 & -9 & 0 & 0 & 0 & 1 & 0 \end{bmatrix}$$

21. Let x_1 = number of Majestic, x_2 = number of Traditional, x_3 = number of Wall clocks.

Maximize $z = 400x_1 + 250x_2 + 160x_3$ subject to
$$4x_1 + 2x_2 + x_3 \le 120 \quad \text{(Cutting)}$$
$$3x_1 + 2x_2 + x_3 \le 80 \quad \text{(Sanding)}$$
$$x_1 + x_2 + .5x_3 \le 40 \quad \text{(Packing)}$$
$$x_1 \ge 0, \ x_2 \ge 0, \ x_3 \ge 0$$

$$\begin{bmatrix} 4 & 2 & 1 & 1 & 0 & 0 & 0 & 120 \\ 3 & 2 & 1 & 0 & 1 & 0 & 0 & 80 \\ 1 & 1 & .5 & 0 & 0 & 1 & 0 & 40 \\ -400 & -250 & -160 & 0 & 0 & 0 & 1 & 0 \end{bmatrix}$$

23. (a) $$3x_1 + 4x_2 + 7x_3 + x_4 + s_1 = 85$$
$$2x_1 + x_2 + 6x_3 + 5x_4 + s_2 = 110$$

(b) Since s_1 is positive and the point $(x_1, x_2, x_3, x_4, s_1, s_2)$ satisfies the first equation, this indicates that the point (x_1, x_2, x_3, x_4) is a solution to the constraint $3x_1 + 4x_2 + 7x_3 + x_4 \le 85$. The point does not lie on the constraint since s_1 is positive.

This point is in the original feasible region if it also satisfies the second constraint. Otherwise it is in the solution half-plane of the first constraint and in the half-plane opposite the solution for the second constraint.

(c) (i) If $s_1 > 0$ and $s_2 > 0$, the point (x, x_2, x_3, x_4) is in the feasible region and is not on either of the boundaries
$$3x_1 + 4x_2 + 7x_3 + x_4 = 85$$
or $$2x_1 + x_2 + 6x_3 + 5x_4 = 110$$
Thus, the point is in the interior of the feasible region.

(ii) $s_1 = 0$ and $s_2 > 0$ indicates (x_1, x_2, x_3, x_4) is on the boundary $3x_1 + 4x_2 + 7x_3 + x_4 = 85$ and is not on the boundary $2x_1 + x_2 + 6x_3 + 5x_4 = 110$ but it is in the feasible region.

(iii) Both $s_1 = 0$ and $s_2 = 0$ indicates the point (x_1, x_2, x_3, x_4) lies in the intersection of the boundaries

$$3x_1 + 4x_2 + 7x_3 + x_4 = 85$$
$$2x_1 + x_2 + 6x_3 + 5x_4 = 110$$

25. Let x_1 = number of pounds of Early Riser

x_2 = number of pounds of Coffee Time

x_3 = number of pounds of After Dinner

x_4 = number of pounds of Deluxe

Maximize profit

$$z = 1.00x_1 + 1.10x_2 + 1.15x_3 + 1.20x_4 \quad \text{subject to}$$
$$.80x_1 + .75x_2 + .75x_3 + .50x_4 \leq 260 \ (\text{regular})$$
$$.20x_1 + .23x_2 + .20x_3 + .40x_4 \leq 90 \ (\text{High Mountain})$$
$$.02x_2 + .05x_3 + .10x_4 \leq 20 \ (\text{Chocolate})$$
$$x_1 \geq 0, \ x_2 \geq 0, \ x_3 \geq 0, \ x_4 \geq 0$$

$$\begin{bmatrix} .80 & .75 & .75 & .50 & 1 & 0 & 0 & 0 & 260 \\ .20 & .23 & .20 & .40 & 0 & 1 & 0 & 0 & 90 \\ 0 & .02 & .05 & .10 & 0 & 0 & 1 & 0 & 20 \\ -1.00 & -1.10 & -1.15 & -1.20 & 0 & 0 & 0 & 1 & 0 \end{bmatrix}$$

27. Let x_1 = amount invested in stocks

x_2 = amount invested in treasury bonds

x_3 = amount invested in municipal bonds

x_4 = amount invested in corporate bonds

The linear programming problem is stated as follows:
Maximize return on investment

$$z = .10x_1 + .06x_2 + .07x_3 + .08x_4 \quad \text{subject to}$$
$$x_1 + x_2 + x_3 + x_4 \leq 10,000 \ (\text{total investment})$$
$$x_1 \leq .40(x_1 + x_2 + x_3 + x_4) \ (\text{stocks less than 40\%})$$
$$x_3 \leq .30(x_1 + x_2 + x_3 + x_4) \ (\text{municipal bonds less than 30\%})$$
$$x_4 \leq .25(x_1 + x_2 + x_3 + x_4) \ (\text{corporate bonds less than 25\%})$$

Simplifying the last four constraints the problem is
Maximize

$$z = .10x_1 + .06x_2 + .07x_3 + .08x_4 \quad \text{subject to}$$
$$.60x_1 - .40x_2 - .40x_3 - .40x_4 \leq 0$$
$$-.30x_1 - .30x_2 + .70x_3 - .30x_4 \leq 0$$
$$-.25x_1 - .25x_2 - .25x_3 + .75x_4 \leq 0$$
$$x_1 \geq 0, \ x_2 \geq 0, \ x_3 \geq 0, \ x_4 \geq 0$$

$$\begin{bmatrix} .60 & -.40 & -.40 & -.40 & 1 & 0 & 0 & 0 & 0 \\ -.30 & -.30 & .70 & -.30 & 0 & 1 & 0 & 0 & 0 \\ -.25 & -.25 & -.25 & .75 & 0 & 0 & 1 & 0 & 0 \\ -.10 & -.06 & -.07 & -.08 & 0 & 0 & 0 & 1 & 0 \end{bmatrix}$$

29.   (a)   Enter the matrix

$$\begin{bmatrix} 6 & 3 & 1 & 0 & 0 & 18 \\ 5 & 2 & 0 & 1 & 0 & 27 \\ -40 & -22 & 0 & 0 & 1 & 0 \end{bmatrix}$$

     (b)   Enter the matrix

$$\begin{bmatrix} 10 & 14 & 1 & 0 & 0 & 0 & 73 \\ 6 & 21 & 0 & 1 & 0 & 0 & 67 \\ 15 & 8 & 0 & 0 & 1 & 0 & 48 \\ -134 & -109 & 0 & 0 & 0 & 1 & 0 \end{bmatrix}$$

     (c)   Enter the matrix

$$\begin{bmatrix} 1 & 1 & 1 & 1 & 0 & 0 & 0 & 24 \\ 3 & 1 & 4 & 0 & 1 & 0 & 0 & 37 \\ 2 & 5 & 3 & 0 & 0 & 1 & 0 & 41 \\ -15 & -23 & -34 & 0 & 0 & 0 & 1 & 0 \end{bmatrix}$$

     (d)   Enter the matrix

$$\begin{bmatrix} 7 & 4 & 1 & 2 & 1 & 0 & 0 & 0 & 435 \\ 5 & 3 & 6 & 1 & 0 & 1 & 0 & 0 & 384 \\ 2 & 8 & 4 & 5 & 0 & 0 & 1 & 0 & 562 \\ -24 & -19 & -15 & -33 & 0 & 0 & 0 & 1 & 0 \end{bmatrix}$$

     (e)   Enter the matrix

$$\begin{bmatrix} 4.7 & 3.2 & 1.58 & 1 & 0 & 0 & 0 & 40.6 \\ 2.14 & 1.82 & 5.09 & 0 & 1 & 0 & 0 & 61.7 \\ 1.63 & 3.44 & 2.84 & 0 & 0 & 1 & 0 & 54.8 \\ -12.9 & -11.27 & -23.85 & 0 & 0 & 0 & 1 & 0 \end{bmatrix}$$

Section 4.2

1.   Basic: $x_1 = 8$, $s_2 = 10$, $z = 14$;   nonbasic: $x_2 = 0$, $s_1 = 0$

3.   Basic: $x_2 = 86$, $s_1 = 54$, $s_3 = 39$, $z = 148$;
        nonbasic: $x_1 = x_3 = s_2 = 0$

5.
$$\begin{bmatrix} 5 & 4 & 3 & 1 & 0 & 0 & 0 & 8 \\ 2 & 7 & 1 & 0 & 1 & 0 & 0 & 15 \\ \underline{6} & \underline{8} & \underline{5} & \underline{0} & \underline{0} & \underline{1} & \underline{0} & \underline{24} \\ -8 & -10 & -4 & 0 & 0 & 0 & 1 & 0 \end{bmatrix} \begin{matrix} 8/4 = 2 \\ 15/7 = 2.14 \\ 24/8 = 3 \\ \end{matrix}$$
4 in row 1, column 2

7.
$$\begin{bmatrix} 2 & 5 & 3 & 1 & 0 & 0 & 0 & 15 \\ 4 & 1 & 4 & 0 & 1 & 0 & 0 & 12 \\ \underline{7} & \underline{3} & \underline{-5} & \underline{0} & \underline{0} & \underline{1} & \underline{0} & \underline{10} \\ -25 & -30 & -50 & 0 & 0 & 0 & 1 & 0 \end{bmatrix} \begin{matrix} 15/3 = 5 \\ 12/4 = 3 \\ 10/-5 = -2 \\ \end{matrix}$$
4 in row 2, column 3

9.
$$\begin{bmatrix} 2 & 1 & 1 & 0 & 0 & 0 & 7 \\ 3 & 4 & 0 & 1 & 0 & 0 & 12 \\ \underline{2} & \underline{5} & \underline{0} & \underline{0} & \underline{1} & \underline{0} & \underline{15} \\ -5 & -8 & 0 & 0 & 0 & 1 & 0 \end{bmatrix} \begin{matrix} 7/1 = 7 \\ 12/4 = 3 \\ 15/5 = 3 \\ \end{matrix}$$
Either the 4 in row 2, column 2 or the 5 in row 3, column 2.

11. $$\begin{bmatrix} 3 & 5 & 6 & 1 & 0 & 0 & 0 & 9 \\ 2 & 8 & 2 & 0 & 1 & 0 & 0 & 6 \\ 5 & 4 & 3 & 0 & 0 & 1 & 0 & 15 \\ -6 & -12 & -12 & 0 & 0 & 0 & 1 & 0 \end{bmatrix} \begin{matrix} 9/6 \text{ or } 9/5 \\ 6/2 \text{ or } 6/8 \\ 15/3 \text{ or } 15/4 \\ \end{matrix}$$

Either the 8 in row 2, column 2 or the 6 in row 1, column 3.

13. $$\begin{bmatrix} 6 & 2 & 1 & 0 & 0 & 0 & 3 \\ 4 & 3 & 0 & 1 & 0 & 0 & 0 \\ 3 & 5 & 0 & 0 & 1 & 0 & 8 \\ -12 & -3 & 0 & 0 & 0 & 1 & 0 \end{bmatrix} \begin{matrix} 3/6 = 1/2 \\ 0/4 = 0 \\ 8/3 \\ \end{matrix}$$

4 in row 2, column 1

15. $$\begin{bmatrix} 2 & 3 & 1 & 0 & 0 & 0 & 12 \\ 1 & 2 & 0 & 1 & 0 & 0 & 6 \\ 2 & 5 & 0 & 0 & 1 & 0 & 20 \\ -4 & -3 & 0 & 0 & 0 & 1 & 0 \end{bmatrix} \qquad \begin{bmatrix} 0 & -1 & 1 & -2 & 0 & 0 & 0 \\ 1 & 2 & 0 & 1 & 0 & 0 & 6 \\ 0 & 1 & 0 & -2 & 1 & 0 & 8 \\ 0 & 5 & 0 & 4 & 0 & 1 & 24 \end{bmatrix}$$

17. $$\begin{bmatrix} 6 & 11 & 4 & 1 & 0 & 0 & 0 & 250 \\ -5 & -14 & -8 & 0 & 1 & 0 & 0 & -460 \\ -1 & -1 & -3 & 0 & 0 & 1 & 0 & -390 \\ -10 & -50 & -30 & 0 & 0 & 0 & 1 & 0 \end{bmatrix}$$

$$\begin{bmatrix} 14/3 & 29/3 & 0 & 1 & 0 & 4/3 & 0 & -270 \\ -7/3 & -34/3 & 0 & 0 & 1 & -8/3 & 0 & 580 \\ 1/3 & 1/3 & 1 & 0 & 0 & -1/3 & 0 & 130 \\ 0 & -40 & 0 & 0 & 0 & -10 & 1 & 3900 \end{bmatrix}$$

19. $$\begin{bmatrix} 3 & 1 & 1 & 0 & 0 & 22 \\ 3 & 4 & 0 & 1 & 0 & 34 \\ -2 & -1 & 0 & 0 & 1 & 0 \end{bmatrix} \begin{matrix} 22/3 \\ 34/3 \\ \end{matrix} \qquad \begin{bmatrix} 1 & 1/3 & 1/3 & 0 & 0 & 22/3 \\ 0 & 3 & -1 & 1 & 0 & 12 \\ 0 & -1/3 & 2/3 & 0 & 1 & 44/3 \end{bmatrix}$$

$$\begin{bmatrix} 1 & 0 & 4/9 & -1/9 & 0 & 6 \\ 0 & 1 & -1/3 & 1/3 & 0 & 4 \\ 0 & 0 & 5/9 & 1/9 & 1 & 16 \end{bmatrix} \qquad x_1 = 6, \ x_2 = 4, \ z = 16$$

21. $$\begin{bmatrix} 1 & 4 & 1 & 0 & 0 & 9 \\ 4 & 1 & 0 & 1 & 0 & 6 \\ -4 & -5 & 0 & 0 & 1 & 0 \end{bmatrix} \begin{matrix} 9/4 = 2.25 \\ 6/1 = 6 \\ \end{matrix} \qquad \begin{bmatrix} 1/4 & 1 & 1/4 & 0 & 0 & 9/4 \\ 15/4 & 0 & -1/4 & 1 & 0 & 15/4 \\ -11/4 & 0 & 5/4 & 0 & 1 & 45/4 \end{bmatrix}$$

$$\begin{bmatrix} 0 & 1 & 4/15 & -1/15 & 0 & 2 \\ 1 & 0 & -1/15 & 4/15 & 0 & 1 \\ 0 & 0 & 16/15 & 11/15 & 1 & 14 \end{bmatrix} \qquad x_1 = 1, \ x_2 = 2, \ z = 14$$

23. $$\begin{bmatrix} 1 & 1 & 1 & 0 & 0 & 240 \\ 4 & 3 & 0 & 1 & 0 & 720 \\ -8 & -4 & 0 & 0 & 1 & 0 \end{bmatrix} \begin{matrix} 240 \\ 180 \\ \end{matrix} \qquad \begin{bmatrix} 0 & 1/4 & 1 & -1/4 & 0 & 60 \\ 1 & 3/4 & 0 & 1/4 & 0 & 180 \\ 0 & 2 & 0 & 2 & 1 & 1440 \end{bmatrix}$$

$$x_1 = 180, \ x_2 = 0, \ z = 1440$$

25.
$$\begin{bmatrix} 5 & 5 & 10 & 1 & 0 & 0 & 0 & | & 1000 \\ 10 & 8 & 5 & 0 & 1 & 0 & 0 & | & 2000 \\ 10 & 5 & 0 & 0 & 0 & 1 & 0 & | & 500 \\ -100 & -200 & -50 & 0 & 0 & 0 & 1 & | & 0 \end{bmatrix}$$

$$\begin{bmatrix} -5 & 0 & 10 & 1 & 0 & -1 & 0 & | & 500 \\ -6 & 0 & 5 & 0 & 1 & -8/5 & 0 & | & 1200 \\ 2 & 1 & 0 & 0 & 0 & 1/5 & 0 & | & 100 \\ 300 & 0 & -50 & 0 & 0 & 40 & 1 & | & 20000 \end{bmatrix}$$

$$\begin{bmatrix} -1/2 & 0 & 1 & 1/10 & 0 & -1/10 & 0 & | & 50 \\ -7/2 & 0 & 0 & -1/2 & 1 & -11/10 & 0 & | & 950 \\ 2 & 1 & 0 & 0 & 0 & 1/5 & 0 & | & 100 \\ 275 & 0 & 0 & 5 & 0 & 35 & 1 & | & 22500 \end{bmatrix}$$

$x_1 = 0$, $x_2 = 100$, $x_3 = 50$, $z = 22{,}500$

27.
$$\begin{bmatrix} 1 & 1 & 1 & 1 & 0 & 0 & 0 & | & 100 \\ 3 & 2 & 4 & 0 & 1 & 0 & 0 & | & 210 \\ 1 & 2 & 0 & 0 & 0 & 1 & 0 & | & 150 \\ -3 & -5 & -5 & 0 & 0 & 0 & 1 & | & 0 \end{bmatrix}$$

$$\begin{bmatrix} 1/4 & 1/2 & 0 & 1 & -1/4 & 0 & 0 & | & 47.5 \\ 3/4 & 1/2 & 1 & 0 & 1/4 & 0 & 0 & | & 52.5 \\ 1 & 2 & 0 & 0 & 0 & 1 & 0 & | & 150 \\ 3/4 & -5/2 & 0 & 0 & 5/4 & 0 & 1 & | & 262.5 \end{bmatrix}$$

$$\begin{bmatrix} 0 & 0 & 0 & 1 & -1/4 & -1/4 & 0 & | & 10 \\ 1/2 & 0 & 1 & 0 & 1/4 & -1/4 & 0 & | & 15 \\ 1/2 & 1 & 0 & 0 & 0 & 1/2 & 0 & | & 75 \\ 2 & 0 & 0 & 0 & 5/4 & 5/4 & 1 & | & 450 \end{bmatrix}$$

$x_1 = 0$, $x_2 = 75$, $x_3 = 15$, $z = 450$

29.
$$\begin{bmatrix} 2 & 1 & 4 & 1 & 0 & 0 & 0 & | & 360 \\ 2 & 5 & 10 & 0 & 1 & 0 & 0 & | & 850 \\ 3 & 3 & 1 & 0 & 0 & 1 & 0 & | & 510 \\ -15 & -9 & -15 & 0 & 0 & 0 & 1 & | & 0 \end{bmatrix}$$

$$\begin{bmatrix} 0 & -1 & 10/3 & 1 & 0 & -2/3 & 0 & | & 20 \\ 0 & 3 & 28/3 & 0 & 1 & -2/3 & 0 & | & 510 \\ 1 & 1 & 1/3 & 0 & 0 & 1/3 & 0 & | & 170 \\ 0 & 6 & -10 & 0 & 0 & 5 & 1 & | & 2550 \end{bmatrix}$$

$$\begin{bmatrix} 0 & -3/10 & 1 & 3/10 & 0 & -1/5 & 0 & | & 6 \\ 0 & 29/5 & 0 & -14/5 & 1 & 6/5 & 0 & | & 454 \\ 1 & 11/10 & 0 & -1/10 & 0 & 2/5 & 0 & | & 168 \\ 0 & 3 & 0 & 3 & 0 & 3 & 1 & | & 2610 \end{bmatrix}$$

$x_1 = 168$, $x_2 = 0$, $x_3 = 6$, $z = 2610$

31.
$$\begin{bmatrix} 1 & 8 & 1 & 0 & 0 & 0 & | & 66 \\ 3 & 9 & 0 & 1 & 0 & 0 & | & 72 \\ 2 & 6 & 0 & 0 & 1 & 0 & | & 48 \\ -33 & -9 & 0 & 0 & 0 & 1 & | & 0 \end{bmatrix} \qquad \begin{bmatrix} 0 & 5 & 1 & 0 & -1/2 & 0 & | & 42 \\ 0 & 0 & 0 & 1 & -3/2 & 0 & | & 0 \\ 1 & 3 & 0 & 0 & 1/2 & 0 & | & 24 \\ 0 & 90 & 0 & 0 & 33/2 & 1 & | & 792 \end{bmatrix}$$

$x_1 = 24$, $x_2 = 0$, $z = 792$

76

33.

$$\left[\begin{array}{ccccccc|c} 2 & 1 & 2 & 1 & 0 & 0 & 0 & 100 \\ 1 & 2 & 2 & 0 & 1 & 0 & 0 & 100 \\ 2 & 2 & 1 & 0 & 0 & 1 & 0 & 100 \\ -22 & -20 & -18 & 0 & 0 & 0 & 1 & 0 \end{array}\right]$$

$$\left[\begin{array}{ccccccc|c} 1 & 1/2 & 1 & 1/2 & 0 & 0 & 0 & 50 \\ 0 & 3/2 & 1 & -1/2 & 1 & 0 & 50 & 50 \\ 0 & 1 & -1 & -1 & 0 & 1 & 0 & 0 \\ 0 & -9 & 4 & 11 & 0 & 0 & 1 & 1100 \end{array}\right]$$

$$\left[\begin{array}{ccccccc|c} 1 & 0 & 3/2 & 1 & 0 & -1/2 & 0 & 50 \\ 0 & 0 & 5/2 & 1 & 1 & -3/2 & 0 & 50 \\ 0 & 1 & -1 & -1 & 0 & 1 & 0 & 0 \\ 0 & 0 & -5 & 2 & 0 & 9 & 1 & 1100 \end{array}\right]$$

$$\left[\begin{array}{ccccccc|c} 1 & 0 & 0 & 2/5 & -3/5 & 2/5 & 0 & 20 \\ 0 & 0 & 1 & 2/5 & 2/5 & -3/5 & 0 & 20 \\ 0 & 1 & 0 & -3/5 & 2/5 & 2/5 & 0 & 20 \\ 0 & 0 & 0 & 4 & 2 & 6 & 1 & 1200 \end{array}\right]$$

$x_1 = 20$, $x_2 = 20$, $x_3 = 20$, $z = 1200$

35.   Let x_1 = number cartons of screwdrivers, x_2 = number cartons of chisels, x_3 = number cartons of putty knives.

$$\left[\begin{array}{cccccc|c} 3 & 4 & 5 & 1 & 0 & 0 & 2200 \\ 15 & 12 & 11 & 0 & 1 & 0 & 8500 \\ -5 & -6 & -5 & 0 & 0 & 1 & 0 \end{array}\right]$$

$$\left[\begin{array}{cccccc|c} 3/4 & 1 & 5/4 & 1/4 & 0 & 0 & 550 \\ 6 & 0 & -4 & -3 & 1 & 0 & 1900 \\ -1/2 & 0 & 5/2 & 3/2 & 0 & 1 & 3300 \end{array}\right]$$

$$\left[\begin{array}{cccccc|c} 0 & 1 & 7/4 & 5/8 & -1/8 & 0 & 312.5 \\ 1 & 0 & -2/3 & -1/2 & 1/6 & 0 & 950/3 \\ 0 & 0 & 13/6 & 5/4 & 1/12 & 1 & 10375/3 \end{array}\right]$$

316.7 cartons of screwdrivers, 312.5 of chisels, and no putty knives, for profit of $3458.33. It is reasonable to round to 317 cartons of screwdrivers, 312 cartons of chisels, and no putty knives.

37.   Let x_1 = number of Pack I, x_2 = number of Pack II, and x_3 = number of Pack III. The limitations imposed by the number of books available are

Short story: $3x_1 + 2x_2 + x_3 \le 660$
Science: $x_1 + 4x_2 + 2x_3 \le 740$
History: $2x_1 + x_2 + 3x_3 \le 853$

The linear programming problem to be solved is

Maximize $z = 1.25_1 + 2.00x_2 + 1.60x_3$ subject to

$$3x_1 + 2x_2 + x_3 \le 660$$
$$x_1 + 4x_2 + 2x_3 \le 740$$
$$2x_1 + x_2 + 3x_3 \le 853$$
$$x_1 \ge 0, \; x_2 \ge 0, \; x_3 \ge 0$$

Section 4.2

The initial simplex tableau is

$$\begin{bmatrix} 3 & 2 & 1 & 1 & 0 & 0 & 0 & 660 \\ 1 & 4 & 2 & 0 & 1 & 0 & 0 & 740 \\ 2 & 1 & 3 & 0 & 0 & 1 & 0 & 853 \\ \hline -1.25 & -2.00 & -1.60 & 0 & 0 & 0 & 1 & 0 \end{bmatrix}$$

Here is the sequence of pivots that lead to the solution.

$$\begin{bmatrix} 5/2 & 0 & 0 & 1 & -1/2 & 0 & 0 & 290 \\ 1/4 & 1 & 1/2 & 0 & 1/4 & 0 & 0 & 185 \\ 7/4 & 0 & 5/2 & 0 & -1/4 & 1 & 0 & 668 \\ \hline -3/4 & 0 & -3/5 & 0 & 1/2 & 0 & 1 & 370 \end{bmatrix}$$

$$\begin{bmatrix} 1 & 0 & 0 & 2/5 & -1/5 & 0 & 0 & 116 \\ 0 & 1 & 1/2 & -1/10 & 3/10 & 0 & 0 & 156 \\ 0 & 0 & 5/2 & -7/10 & 1/10 & 1 & 0 & 465 \\ \hline 0 & 0 & -3/5 & 3/10 & 7/20 & 0 & 1 & 457 \end{bmatrix}$$

$$\begin{bmatrix} 1 & 0 & 0 & 2/5 & -1/5 & 0 & 0 & 116 \\ 0 & 1 & 0 & 1/25 & 7/25 & -1/5 & 0 & 63 \\ 0 & 0 & 1 & -7/25 & 1/25 & 2/5 & 0 & 186 \\ \hline 0 & 0 & 0 & 33/250 & 187/500 & 6/25 & 1 & 568.6 \end{bmatrix}$$

The maximum profit is $568.60 using 116 of Pack I, 63 of Pack II, and 186 of Pack III.

39. Let x_1 = lbs of Early Riser, x_2 = lbs of After Dinner, x_3 = lbs of Deluxe.

Maximize $z = x_1 + 1.1x_2 + 1.2x_3$, subject to
$$.80x_1 + .75x_2 + .50x_3 \le 255$$
$$.20x_1 + .20x_2 + .40x_3 \le 80$$
$$.05x_2 + .10x_3 \le 15$$
$$x_1 \ge 0,\ x_2 \ge 0,\ x_3 \ge 0$$

$$\begin{bmatrix} .80 & .75 & .50 & 1 & 0 & 0 & 0 & 255 \\ .20 & .20 & .40 & 0 & 1 & 0 & 0 & 80 \\ 0 & .05 & .10 & 0 & 0 & 1 & 0 & 15 \\ \hline -1 & -1.1 & -1.2 & 0 & 0 & 0 & 1 & 0 \end{bmatrix}$$

$$\begin{bmatrix} 4/5 & 1/2 & 0 & 1 & 0 & -5 & 0 & 180 \\ 1/5 & 0 & 0 & 0 & 1 & -4 & 0 & 20 \\ 0 & 1/2 & 1 & 0 & 0 & 10 & 0 & 150 \\ \hline -1 & -1/2 & 0 & 0 & 0 & 12 & 1 & 180 \end{bmatrix}$$

$$\begin{bmatrix} 0 & 1/2 & 0 & 1 & -4 & 11 & 0 & 100 \\ 1 & 0 & 0 & 0 & 5 & -20 & 0 & 100 \\ 0 & 1/2 & 1 & 0 & 0 & 10 & 0 & 150 \\ \hline 0 & -1/2 & 0 & 0 & 5 & -8 & 1 & 280 \end{bmatrix}$$

$$\begin{bmatrix} 0 & 1 & 0 & 2 & -8 & 22 & 0 & 200 \\ 1 & 0 & 0 & 0 & 5 & -20 & 0 & 100 \\ 0 & 0 & 1 & -1 & 4 & -1 & 0 & 50 \\ \hline 0 & 0 & 0 & 1 & 1 & 3 & 1 & 380 \end{bmatrix}$$

Maximum profit is $380 using 100 lbs Early Riser, 200 lbs After Dinner, and 50 lbs Deluxe.

41. Let x_1 = number of packages of TV mix, x_2 = number of packages of Party Mix, x_3 = number of packages of Dinner Mix. Maximize $z = 4.4x_1 + 4.8x_2 + 5.2x_3$, subject to

$$600x_1 + 500x_2 + 300x_3 \leq 39{,}500$$
$$300x_1 + 300x_2 + 200x_3 \leq 22{,}500$$
$$100x_1 + 200x_2 + 400x_3 \leq 16{,}500$$
$$x_1 \geq 0,\ x_2 \geq 0,\ x_3 \geq 0$$

$$\begin{bmatrix} 600 & 500 & 300 & 1 & 0 & 0 & 0 & 39500 \\ 300 & 300 & 200 & 0 & 1 & 0 & 0 & 22500 \\ 100 & 200 & 400 & 0 & 0 & 1 & 0 & 16500 \\ -4.4 & -4.8 & -5.2 & 0 & 0 & 0 & 1 & 0 \end{bmatrix}$$

$$\begin{bmatrix} 525 & 350 & 0 & 1 & 0 & -3/4 & & 27125 \\ 250 & 200 & 0 & 0 & 1 & -1/2 & & 14250 \\ 1/4 & 1/2 & 1 & 0 & 0 & 1/400 & & 165/4 \\ -31/10 & -11/5 & 0 & 0 & 0 & 13/1000 & & 429/2 \end{bmatrix}$$

$$\begin{bmatrix} 1 & 2/3 & 0 & 1/525 & 0 & -1/700 & 0 & 155/3 \\ 0 & 100/3 & 0 & -10/21 & 1 & -1/7 & 0 & 4000/3 \\ 0 & 1/3 & 1 & -1/2100 & 0 & 1/350 & 0 & 85/3 \\ 0 & -2/15 & 0 & 31/5250 & 0 & 3/350 & 1 & 1124/3 \end{bmatrix}$$

$$\begin{bmatrix} 1 & 0 & 0 & 2/175 & -1/50 & 1/700 & 0 & 25 \\ 0 & 1 & 0 & -1/70 & 3/100 & -3/700 & 0 & 40 \\ 0 & 0 & 1 & 3/700 & -1/100 & 3/700 & 0 & 15 \\ 0 & 0 & 0 & 1/250 & 1/250 & 1/125 & 1 & 380 \end{bmatrix}$$

Maximum revenue is $380 making 25 packages of TV Mix, 40 packages of Party Mix, and 15 packages of Dinner Mix.

43. Let x = number of Majestic, y = number of Traditional, z = number of Wall clocks. Maximize $R = 400x + 250y + 160z$, subject to

$$4x + 2y + z \leq 124$$
$$3x + y + z \leq 81$$
$$x + y + .5z \leq 46$$

$$\begin{bmatrix} 4 & 2 & 1 & 1 & 0 & 0 & 0 & 124 \\ 3 & 1 & 1 & 0 & 1 & 0 & 0 & 81 \\ 1 & 1 & .5 & 0 & 0 & 1 & 0 & 46 \\ -400 & -250 & -160 & 0 & 0 & 0 & 1 & 0 \end{bmatrix}$$

$$\begin{bmatrix} 0 & 2/3 & -1/3 & 1 & -4/3 & 0 & 0 & 16 \\ 1 & 1/3 & 1/3 & 0 & 1/3 & 0 & 0 & 27 \\ 0 & 2/3 & 1/6 & 0 & -1/3 & 1 & 0 & 19 \\ 0 & -350/3 & -80/3 & 0 & 400/3 & 0 & 1 & 10800 \end{bmatrix}$$

$$\begin{bmatrix} 0 & 1 & -1/2 & 3/2 & -2 & 0 & 0 & 24 \\ 1 & 0 & 1/2 & -1/2 & 1 & 0 & 0 & 19 \\ 0 & 0 & 1/2 & -1 & 1 & 1 & 0 & 3 \\ 0 & 0 & -85 & 175 & -100 & 0 & 1 & 13600 \end{bmatrix}$$

$$\begin{bmatrix} 0 & 1 & 0 & 1/2 & -1 & 1 & 0 & | & 27 \\ 1 & 0 & 0 & 1/2 & 0 & -1 & 0 & | & 16 \\ 0 & 0 & 1 & -2 & 2 & 2 & 0 & | & 6 \\ \hline 0 & 0 & 0 & 5 & 70 & 170 & 1 & | & 14110 \end{bmatrix}$$

Maximum revenue is $14,110 for 16 Majestic, 27 Traditional, and 6 Wall clocks.

45. The initial tableau and ratios using the pivot column, column 1 are

$$\begin{matrix} & & & & & & & & \text{Ratios} \\ \begin{bmatrix} 2 & 1 & 1 & 0 & 0 & 0 & | & 80 \\ 2 & 3 & 0 & 1 & 0 & 0 & | & 120 \\ 4 & 1 & 0 & 0 & 1 & 0 & | & 160 \\ \hline -5 & -4 & 0 & 0 & 0 & 1 & | & 0 \end{bmatrix} & \begin{matrix} 40 \\ 60 \\ 40 \\ \end{matrix} \end{matrix}$$

Either row 1 or row 3 may be used as the pivot row. Let's use row 1 and we obtain the next tableau

$$\begin{matrix} & & & & & & & & \text{Ratios} \\ \begin{bmatrix} 1 & 1/2 & 1/2 & 0 & 0 & 0 & | & 40 \\ 0 & 2 & -1 & 1 & 0 & 0 & | & 40 \\ 0 & -1 & -2 & 0 & 1 & 0 & | & 0 \\ \hline 0 & -3/2 & 5/2 & 0 & 0 & 1 & | & 200 \end{bmatrix} & \begin{matrix} 80 \\ 20 \\ -0 \\ \end{matrix} \end{matrix}$$

This is not optimal so we use column 2 as the next pivot column. The smallest ratio 0/(-1) has a negative divisor so we choose the least positive divisor 20, to determine that row 2 is the pivot row. Pivoting on 2 in row 2, column 2, we obtain

$$\begin{bmatrix} 1 & 0 & 3/4 & -1/4 & 0 & 0 & | & 30 \\ 0 & 1 & -1/2 & 1/2 & 0 & 0 & | & 20 \\ 0 & 0 & -5/2 & 1/2 & 1 & 0 & | & 20 \\ \hline 0 & 0 & 7/4 & 3/4 & 0 & 1 & | & 230 \end{bmatrix}$$

This yields maximum z = 230 at (30, 20). The sequence of corner points was (0, 0) (40, 0), and (30, 20).

Now go back to the initial tableau and pivot on row 3, column 1. We obtain

$$\begin{matrix} & & & & & & & & \text{Ratios} \\ \begin{bmatrix} 0 & 1/2 & 1 & 0 & -1/2 & 0 & | & 0 \\ 0 & 5/2 & 0 & 1 & -1/2 & 0 & | & 40 \\ 1 & 1/4 & 0 & 0 & 1/4 & 0 & | & 40 \\ \hline 0 & -11/4 & 0 & 0 & 5/4 & 1 & | & 200 \end{bmatrix} & \begin{matrix} +0 \\ 16 \\ 160 \\ \end{matrix} \end{matrix}$$

The zero ratio has a positive divisor so it determines the pivot row, row 1. A pivot on 1/2 in row 1, column 2 gives the tableau

$$\begin{matrix} & & & & & & & & \text{Ratios} \\ \begin{bmatrix} 0 & 1 & 2 & 0 & -1 & 0 & | & 0 \\ 0 & 0 & -5 & 1 & 2 & 0 & | & 40 \\ 1 & 0 & -1/2 & 0 & 1/2 & 0 & | & 40 \\ \hline 0 & 0 & 11/2 & 0 & -3/2 & 1 & | & 200 \end{bmatrix} & \begin{matrix} -0 \\ 20 \\ 80 \\ \end{matrix} \end{matrix}$$

We still need to pivot using column 5 and row 2 since the zero ratio has a negative divisor. That pivot yields

$$\begin{bmatrix} 0 & 1 & -1/2 & 1/2 & 0 & 0 & | & 20 \\ 0 & 0 & -5/2 & 1/2 & 1 & 0 & | & 20 \\ 1 & 0 & 3/4 & -1/4 & 0 & 0 & | & 30 \\ 0 & 0 & 7/4 & 3/4 & 0 & 1 & | & 230 \end{bmatrix}$$

This tableau gives the optimal solution, maximum z = 230 at (30, 20). The sequence of corner points was (0, 0), (40, 0), (40,0), and (30, 20). Notice that this sequence uses the corner (40, 0) twice, once for each constraint passing through (40, 0).

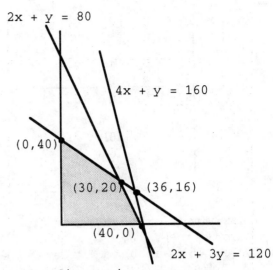

47. The initial tableau and ratios for the first pivot row are

								Ratios
1	2	1	0	0	0	0	100	50
1	1	0	1	0	0	0	60	60
4	5	0	0	1	0	0	265	53
3	5	0	0	0	1	0	255	51
-7	-8	0	0	0	0	1	0	

The sequence of tableaux are

								Ratios
1/2	1	1/2	0	0	0	0	50	100
1/2	0	-1/2	1	0	0	0	10	20
3/2	0	-5/2	0	1	0	0	15	10
1/2	0	-5/2	0	0	1	0	5	10
-3	0	4	0	0	0	1	400	

Use row 3 for the pivot row.

								Ratios
0	1	4/3	0	-1/3	0	0	45	33.75
0	0	1/3	1	-1/3	0	0	5	15
1	0	-5/3	0	2/3	0	0	10	-6
0	0	-5/3	0	-1/3	1	0	0	-0
0	0	-1	0	2	0	1	430	

0	1	0	-4	1	0	0	25
0	0	1	3	-1	0	0	15
1	0	0	5	-1	0	0	35
0	0	0	5	-2	1	0	25
0	0	0	3	1	0	1	445

Maximum z = 445 at (35, 25). The sequence of corner points used was (0, 0), (0, 50), (10, 45), and (35, 25).

Now go back to the second tableau and use row 4 instead of row 3 for the pivot row. We obtain the following sequence.

								Ratios
0	1	3	0	0	-1	0	45	15
0	0	2	1	0	-1	0	5	2.5
0	0	5	0	1	-3	0	0	0
1	0	-5	0	0	2	0	10	-2
0	0	-11	0	6	6	1	430	

81

Ratios

$$\begin{bmatrix} 0 & 1 & 0 & 0 & -3/5 & 4/5 & 0 & 45 \\ 0 & 0 & 0 & 1 & -2/5 & 1/5 & 0 & 5 \\ 0 & 0 & 1 & 0 & 1/5 & -3/5 & 0 & 0 \\ 1 & 0 & 0 & 0 & 1 & -1 & 0 & 10 \\ 0 & 0 & 0 & 0 & 11/5 & -3/5 & 1 & 430 \end{bmatrix}$$

56.25
25
–0
–10

$$\begin{bmatrix} 0 & 1 & 0 & -4 & 1 & 0 & 0 & 25 \\ 0 & 0 & 0 & 5 & -2 & 1 & 0 & 25 \\ 0 & 0 & 1 & 3 & -1 & 0 & 0 & 15 \\ 1 & 0 & 0 & 5 & -1 & 0 & 0 & 35 \\ 0 & 0 & 0 & 3 & 1 & 0 & 1 & 445 \end{bmatrix}$$

Maximum z = 445 at (35, 25). The sequence of corner points used was (0, 0), (0, 50), (10, 45), (10, 45) and (35, 25). The corner (10, 45) was used twice where two constraints intersect.

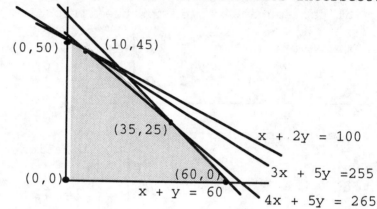

49. Pivot on 1.5 in the (1, 3) position.

$$\begin{bmatrix} 1 & 0 & 1.5 & 1 & 0 & -.5 & 0 & 50 \\ 0 & 0 & 2.5 & 1 & 1 & -1.5 & 0 & 75 \\ 0 & 1 & -1 & -1 & 0 & 1 & 0 & 25 \\ 0 & 0 & -5 & 2 & 0 & 9 & 1 & 1100 \end{bmatrix}$$

$$\begin{bmatrix} .67 & 0 & 1 & .67 & 0 & -.33 & 0 & 33.33 \\ -1.67 & 0 & 0 & -.67 & 1 & -.67 & 0 & -8.33 \\ .67 & 1 & 0 & -.33 & 0 & .67 & 0 & 58.33 \\ 3.33 & 0 & 0 & 5.33 & 0 & 7.33 & 1 & 1266.67 \end{bmatrix}$$

51. Pivot on the (2, 3) entry.

$$\begin{bmatrix} .65 & 1 & .60 & 0 & -.05 & 0 & 0 & 0 & 19.2 \\ -.80 & 0 & 4.80 & 1 & .60 & 0 & 0 & 0 & 81.6 \\ 2.40 & 0 & 8.60 & 0 & .20 & 1 & 0 & 0 & 91.2 \\ 5.6 & 0 & -11.60 & 0 & -1.20 & 0 & 0 & 1 & 460.8 \end{bmatrix}$$

$$\begin{bmatrix} .75 & 1 & 0 & -.13 & -.13 & 0 & 0 & 0 & 9 \\ -.17 & 0 & 1 & .21 & .13 & 0 & 0 & 0 & 17 \\ 3.83 & 0 & 0 & -1.79 & -.88 & 1 & 0 & 0 & -55 \\ 3.67 & 0 & 0 & 2.42 & .25 & 0 & 0 & 1 & 658 \end{bmatrix}$$

The entries may vary some from those shown depending on how you round the numbers.

53. The initial tableau is

$$\begin{bmatrix} 40 & 25 & 30 & 45 & 1 & 0 & 1 & 0 & 1250 \\ 25 & 50 & 40 & 45 & 0 & 1 & 0 & 0 & 1600 \\ 6 & 4 & 8 & 2 & 0 & 0 & 1 & 0 & 195 \\ -48 & -45 & -55 & -65 & 0 & 0 & 0 & 1 & 0 \end{bmatrix}$$

The final tableau is

$$\begin{bmatrix} .74 & 0 & 0 & 1 & .04 & -.01 & -.08 & 0 & 12.02 \\ -1.03 & 1 & 0 & 0 & -.05 & .05 & -.07 & 0 & 6.81 \\ 1.08 & 0 & 1 & 0 & .01 & -.02 & .18 & 0 & 17.96 \\ 13.21 & 0 & 0 & 0 & 1.19 & .19 & 1.47 & 1 & 2075.71 \end{bmatrix}$$

The solution is (0, 6.81, 17.96, 12.02) with a profit of $2075.71. However, the solution must be an integer number of patterns, so we might round the solution to (0, 7, 18, 12) which gives a maximum profit of $2085. Unfortunately, this requires 1255 tulips, 1610 daffodils, and 196 boxwood while there are only 1250 tulips, 1600 daffodils and 195 boxwood available. We can round down to (0, 6, 18, 12) with a profit of $2040 and remain within the constraints.

Section 4.3

1. $\begin{bmatrix} 2 & 4 \\ 1 & 0 \\ 3 & 2 \end{bmatrix}$

3. $\begin{bmatrix} 4 & 1 & 6 & 2 \\ 3 & 8 & -7 & 4 \\ 2 & -2 & 1 & 6 \end{bmatrix}$

5. (a) $\begin{bmatrix} 6 & 5 & 30 \\ 8 & 3 & 42 \\ 25 & 30 & 1 \end{bmatrix}$

    (b) $\begin{bmatrix} 6 & 8 & 25 \\ 5 & 3 & 30 \\ 30 & 42 & 1 \end{bmatrix}$

    (c) $\begin{bmatrix} 6 & 8 & 1 & 0 & 0 & 25 \\ 5 & 3 & 0 & 1 & 0 & 30 \\ -30 & -42 & 0 & 0 & 1 & 0 \end{bmatrix}$

7. (a) $\begin{bmatrix} 22 & 30 & 110 \\ 15 & 40 & 95 \\ 20 & 35 & 68 \\ 500 & 700 & 1 \end{bmatrix}$

    (b) $\begin{bmatrix} 22 & 15 & 20 & 500 \\ 30 & 40 & 35 & 700 \\ 110 & 95 & 68 & 1 \end{bmatrix}$

    (c) $\begin{bmatrix} 22 & 15 & 20 & 1 & 0 & 0 & 500 \\ 30 & 40 & 35 & 0 & 1 & 0 & 700 \\ -110 & -95 & -68 & 0 & 0 & 1 & 0 \end{bmatrix}$

9.

y_1	y_2	x_1	x_2	w	
0	1	1	-1	0	1
1	0	-1	2	0	2
0	0	6	4	1	40

From the last row $x_1 = 6$, $x_2 = 4$, $z = 40$.

11.

y_1	y_2	y_3	x_1	x_2	x_3	w	
1/6	0	1	1/5	-1/8	0	0	7/6
1/3	1	0	-1/5	1/4	0	0	4/3
-1	0	0	-2/5	-1	1	0	1
45	0	0	12	10	0	1	510

Minimum z = 510 at (12,10,0)

13.

$$
\begin{array}{ccccc}
y_1 & y_2 & x_1 & x_2 & w \\
\end{array}
$$

$$
\left[\begin{array}{ccccc|c}
1 & 2 & 1 & 0 & 0 & 4 \\
1 & 1 & 0 & 1 & 0 & 3 \\
\hline
-8 & -14 & 0 & 0 & 1 & 0
\end{array}\right]
\qquad
\begin{array}{ccccc}
y_1 & y_2 & x_1 & x_2 & w \\
\end{array}
\left[\begin{array}{ccccc|c}
1/2 & 1 & 1/2 & 0 & 0 & 2 \\
1/2 & 0 & -1/2 & 1 & 0 & 1 \\
\hline
-1 & 0 & 7 & 0 & 1 & 28
\end{array}\right]
$$

$$
\begin{array}{ccccc}
y_1 & y_2 & x_1 & x_2 & w \\
\end{array}
$$

$$
\left[\begin{array}{ccccc|c}
0 & 1 & 1 & -1 & 0 & 1 \\
1 & 0 & -1 & 2 & 0 & 2 \\
\hline
0 & 0 & 6 & 2 & 1 & 30
\end{array}\right]
\qquad \text{Minimum } z = 30 \text{ at } (6, 2)
$$

15.

$$
\begin{array}{ccccccc}
y_1 & y_2 & y_3 & x_1 & x_2 & x_3 & w
\end{array}
$$

$$
\left[\begin{array}{ccccccc|c}
3 & 1 & 0 & 1 & 0 & 0 & 0 & 10 \\
1 & 1 & 4 & 0 & 1 & 0 & 0 & 16 \\
6 & 0 & 1 & 0 & 0 & 1 & 0 & 20 \\
\hline
-9 & -9 & -12 & 0 & 0 & 0 & 1 & 0
\end{array}\right]
$$

$$
\begin{array}{ccccccc}
y_1 & y_2 & y_3 & x_1 & x_2 & x_3 & w
\end{array}
$$

$$
\left[\begin{array}{ccccccc|c}
3 & 1 & 0 & 1 & 0 & 0 & 0 & 10 \\
1/4 & 1/4 & 1 & 0 & 1/4 & 0 & 0 & 4 \\
23/4 & -1/4 & 0 & 0 & -1/4 & 1 & 0 & 16 \\
\hline
-6 & -6 & 0 & 0 & 3 & 0 & 1 & 48
\end{array}\right]
$$

$$
\begin{array}{ccccccc}
y_1 & y_2 & y_3 & x_1 & x_2 & x_3 & w
\end{array}
$$

$$
\left[\begin{array}{ccccccc|c}
3 & 1 & 0 & 1 & 0 & 0 & 0 & 10 \\
-1/2 & 0 & 1 & -1/4 & 1/4 & 0 & 0 & 3/2 \\
13/2 & 0 & 0 & 1/4 & -1/4 & 1 & 0 & 37/2 \\
\hline
12 & 0 & 0 & 6 & 3 & 0 & 1 & 108
\end{array}\right]
\qquad \text{Minimum } z = 108 \text{ at } (6, 3, 0)
$$

17.

$$
\begin{array}{ccccccc}
y_1 & y_2 & y_3 & x_1 & x_2 & x_3 & w
\end{array}
$$

$$
\left[\begin{array}{ccccccc|c}
1 & 3 & 3 & 1 & 0 & 0 & 0 & 8 \\
1 & 1 & 6 & 0 & 1 & 0 & 0 & 5 \\
1 & 3 & 8 & 0 & 0 & 1 & 0 & 12 \\
\hline
-37 & -81 & -216 & 0 & 0 & 0 & 1 & 0
\end{array}\right]
$$

$$
\begin{array}{ccccccc}
y_1 & y_2 & y_3 & x_1 & x_2 & x_3 & w
\end{array}
$$

$$
\left[\begin{array}{ccccccc|c}
1/2 & 5/2 & 0 & 1 & -1/2 & 0 & 0 & 11/2 \\
1/6 & 1/6 & 1 & 0 & 1/6 & 0 & 0 & 5/6 \\
-1/3 & 5/3 & 0 & 0 & -4/3 & 1 & 0 & 16/3 \\
\hline
-1 & -45 & 0 & 0 & 36 & 0 & 1 & 180
\end{array}\right]
$$

$$
\begin{array}{ccccccc}
y_1 & y_2 & y_3 & x_1 & x_2 & x_3 & w
\end{array}
$$

$$
\left[\begin{array}{ccccccc|c}
1/5 & 1 & 0 & 2/5 & -1/5 & 0 & 0 & 11/5 \\
2/15 & 0 & 1 & -1/15 & 1/5 & 0 & 0 & 7/5 \\
-2/3 & 0 & 0 & -2/3 & -1 & 1 & 0 & 5/3 \\
\hline
8 & 0 & 0 & 18 & 27 & 0 & 1 & 279
\end{array}\right]
\qquad
\begin{array}{l}
\text{Minimum } z = 279 \text{ at} \\
(18, 27, 0)
\end{array}
$$

19. Let x_1 = number at Dallas, x_2 = number at New Orleans.
Minimize $z = 22{,}000x_1 + 12{,}000x_2$
subject to $800x_1 + 500x_2 \geq 28{,}000$
$280x_1 + 150x_2 \geq 9{,}000$
$x_1 \geq 0,\ x_2 \geq 0$

$$
\left[\begin{array}{ccccc|c}
800 & 280 & 1 & 0 & 0 & 22000 \\
500 & 150 & 0 & 1 & 0 & 12000 \\
\hline
-28000 & -9000 & 0 & 0 & 1 & 0
\end{array}\right]
\qquad
\left[\begin{array}{ccccc|c}
0 & 40 & 1 & -8/5 & 0 & 2800 \\
1 & 3/10 & 0 & 1/500 & 0 & 24 \\
\hline
0 & -600 & 0 & 56 & 1 & 672000
\end{array}\right]
$$

$$
\left[\begin{array}{ccccc|c}
0 & 1 & 1/40 & -1/25 & 0 & 70 \\
1 & 0 & -3/400 & 7/500 & 0 & 3 \\
\hline
0 & 0 & 15 & 32 & 1 & 714000
\end{array}\right]
\qquad
\begin{array}{l}
\text{15 at Dallas, 32 at New} \\
\text{Orleans for cost of \$714,000}
\end{array}
$$

84

21. Let x_1 = number of days the Chicago plant operates
     x_2 = number of days the Detroit plant operates.
     Minimize $z = 20{,}000x_1 + 15{,}000x_2$
     subject to $600x_1 + 300x_2 \geq 24{,}000$ (Radial)
                $100x_1 + 100x_2 \geq 5{,}000$ (Standard)
                $x_1 \geq 0,\ x_2 \geq 0.$
     This is a standard minimization problem so we solve it using its
     dual problem. The augmented matrix of the problem is

$$\begin{bmatrix} 600 & 300 & 24000 \\ 100 & 100 & 5000 \\ 20000 & 15000 & 0 \end{bmatrix}$$

     The transpose of A is

$$\begin{bmatrix} 600 & 100 & 20000 \\ 300 & 100 & 15000 \\ 24000 & 5000 & 0 \end{bmatrix}$$ This matrix represents the dual problem.

     The tableaux that solve this dual problem are

$$\begin{bmatrix} 600 & 100 & 1 & 0 & 0 & 20000 \\ 300 & 100 & 0 & 1 & 0 & 15000 \\ -24000 & -5000 & 0 & 0 & 1 & 0 \end{bmatrix}$$

$$\begin{bmatrix} 1 & 1/6 & 1/600 & 0 & 0 & 200/6 \\ 300 & 100 & 0 & 1 & 0 & 15000 \\ -24000 & -5000 & 0 & 0 & 1 & 0 \end{bmatrix}$$

$$\begin{bmatrix} 1 & 1/6 & 1/600 & 0 & 0 & 200/6 \\ 0 & 50 & -1/2 & 1 & 0 & 5000 \\ 0 & -1000 & 40 & 0 & 1 & 800000 \end{bmatrix}$$

$$\begin{bmatrix} 1 & 1/6 & 1/600 & 0 & 0 & 200/6 \\ 0 & 1 & -1/100 & 1/50 & 0 & 100 \\ 0 & -1000 & 40 & 0 & 1 & 800000 \end{bmatrix}$$

$$\begin{bmatrix} 1 & 0 & 1/300 & -1/300 & 0 & 100/6 \\ 0 & 1 & -1/100 & 1/50 & 0 & 100 \\ 0 & 0 & 30 & 20 & 1 & 900000 \end{bmatrix}$$

     The last row gives the solution to the original problem
          $x_1 = 30,\ x_2 = 20,\ z = 900{,}000.$
     The minimum operating costs are $900,000 when the Chicago plant
     operates 30 days and the Detroit plant 20 days.

Section 4.4

1. Change minimum $z = 2x_1 - 5x_2$ to maximum $w = -2x_1 + 5x_2$,

$$\begin{bmatrix} 4 & 3 & 1 & 0 & 0 & 120 \\ 2 & 1 & 0 & 1 & 0 & 50 \\ 2 & -5 & 0 & 0 & 1 & 0 \end{bmatrix} \qquad \begin{bmatrix} 4/3 & 1 & 1/3 & 0 & 0 & 40 \\ 2/3 & 0 & -1/3 & 1 & 0 & 10 \\ 26/3 & 0 & 5/3 & 0 & 1 & 200 \end{bmatrix}$$

    Minimum z is -200 at $x_1 = 0,\ x_2 = 40$

3.
$$\begin{bmatrix} 3 & 2 & -12 & 1 & 0 & 0 & 0 & 120 \\ 2 & 4 & 6 & 0 & 1 & 0 & 0 & 120 \\ 1 & -2 & 3 & 0 & 0 & 1 & 0 & 52 \\ 4 & 5 & -9 & 0 & 0 & 0 & 1 & 0 \end{bmatrix}$$

$$\begin{bmatrix} 7 & -6 & 0 & 1 & 0 & 4 & 0 & 328 \\ 0 & 8 & 0 & 0 & 1 & -2 & 0 & 16 \\ 1/3 & -2/3 & 1 & 0 & 0 & 1/3 & 0 & 52/3 \\ 7 & -1 & 0 & 0 & 0 & 3 & 1 & 156 \end{bmatrix}$$

$$\begin{bmatrix} 7 & 0 & 0 & 1 & 3/4 & 5/2 & 0 & 340 \\ 0 & 1 & 0 & 0 & 1/8 & -1/4 & 0 & 2 \\ 1/3 & 0 & 1 & 0 & 1/12 & 1/6 & 0 & 56/3 \\ 7 & 0 & 0 & 0 & 1/8 & 11/4 & 1 & 158 \end{bmatrix}$$

Minimum $z = -158$ at $x_1 = 0$, $x_2 = 2$, $x_3 = 56/3$

5. Modify the last two constraints to
$$-3x_1 - 5x_2 - 8x_3 \le -106$$
$$-6x_1 - 12x_2 - x_3 \le -98$$
The initial simplex tableau then is

$$\begin{bmatrix} 9 & 7 & 10 & 1 & 0 & 0 & 0 & 154 \\ -3 & -5 & -8 & 0 & 1 & 0 & 0 & -106 \\ -6 & -12 & -1 & 0 & 0 & 1 & 0 & -98 \\ -5 & -3 & -8 & 0 & 0 & 0 & 1 & 0 \end{bmatrix}$$

7. Change the objective function to maximize $w = -7x_1 - 5x_2 - 8x_3$ and change the second and third constraints to
$$-7x_1 - 9x_2 - 15x_3 \le -48$$
$$x_1 + 3x_2 + 5x_3 \le 27$$
$$-x_1 - 3x_2 - 5x_3 \le -27$$
The initial simplex tableau then is

$$\begin{bmatrix} 15 & 23 & 9 & 1 & 0 & 0 & 0 & 0 & 85 \\ -7 & -9 & -15 & 0 & 1 & 0 & 0 & 0 & -48 \\ 1 & 3 & 5 & 0 & 0 & 1 & 0 & 0 & 27 \\ -1 & -3 & -5 & 0 & 0 & 0 & 1 & 0 & -27 \\ 7 & 5 & 8 & 0 & 0 & 0 & 0 & 1 & 0 \end{bmatrix}$$

9.
$$\begin{bmatrix} 3 & 2 & 1 & 0 & 0 & 36 \\ -2 & 1 & 0 & 1 & 0 & -3 \\ -5 & -2 & 0 & 0 & 1 & 0 \end{bmatrix} \qquad \begin{bmatrix} 0 & 7/2 & 1 & 3/2 & 0 & 63/2 \\ 1 & -1/2 & 0 & -1/2 & 0 & 3/2 \\ 0 & -9/2 & 0 & -5/2 & 1 & 15/2 \end{bmatrix}$$

$$\begin{bmatrix} 0 & 1 & 2/7 & 3/7 & 0 & 9 \\ 1 & 0 & 1/7 & -2/7 & 0 & 6 \\ 0 & 0 & 9/7 & -4/7 & 1 & 48 \end{bmatrix} \qquad \begin{bmatrix} 0 & 7/3 & 2/3 & 1 & 0 & 21 \\ 1 & 2/3 & 1/3 & 0 & 0 & 12 \\ 0 & 4/3 & 5/3 & 0 & 1 & 60 \end{bmatrix}$$

Maximum $z = 60$ at $(12, 0)$

11.
$$\begin{bmatrix} 5 & 11 & 1 & 0 & 0 & 350 \\ -15 & -8 & 0 & 1 & 0 & -300 \\ -15 & -22 & 0 & 0 & 1 & 0 \end{bmatrix} \qquad \begin{bmatrix} 0 & 25/3 & 1 & 1/3 & 0 & 250 \\ 1 & 8/15 & 0 & -1/15 & 0 & 20 \\ 0 & -14 & 0 & -1 & 1 & 300 \end{bmatrix}$$

$$\begin{bmatrix} 0 & 1 & 3/25 & 1/25 & 0 & 30 \\ 1 & 0 & -8/125 & -11/125 & 0 & 4 \\ 0 & 0 & 42/25 & -11/25 & 1 & 720 \end{bmatrix}$$

$$\begin{bmatrix} 0 & 25 & 3 & 1 & 0 & 750 \\ 1 & 11/5 & 1/5 & 0 & 0 & 70 \\ 0 & 11 & 3 & 0 & 1 & 1050 \end{bmatrix}$$

Maximum is $z = 1050$ at $x_1 = 70$, $x_2 = 0$

13. Replace the second constraint with $-3x_1 - 6x_2 \leq -120$ and we obtain the initial simplex tableau

$$\begin{bmatrix} 5 & 8 & 1 & 0 & 0 & | & 180 \\ -3 & -6 & 0 & 1 & 0 & | & -120 \\ -11 & -20 & 0 & 0 & 1 & | & 0 \end{bmatrix}$$

We pivot on -6 because of the -120 in the last column. The next tableau is

$$\begin{bmatrix} 1 & 0 & 1 & 4/3 & 0 & | & 20 \\ 1/2 & 1 & 0 & -1/6 & 0 & | & 20 \\ -1 & 0 & 0 & -10/3 & 1 & | & 400 \end{bmatrix}$$

This is not optimal so we choose column 4 as the pivot column, row 1 as the pivot row, and pivot on $4/3$ to obtain

$$\begin{bmatrix} 3/4 & 0 & 3/4 & 1 & 0 & | & 15 \\ 5/8 & 1 & 1/8 & 0 & 0 & | & 45/2 \\ 3/2 & 0 & 5/2 & 0 & 1 & | & 450 \end{bmatrix}$$

This is optimal with $z = 450$ at $(0, 22.5)$.

15.

$$\begin{bmatrix} 6 & 12 & 4 & 1 & 0 & 0 & 0 & | & 900 \\ -5 & -16 & -8 & 0 & 1 & 0 & 0 & | & -120 \\ -3 & -1 & -1 & 0 & 0 & 1 & 0 & | & -300 \\ -10 & -50 & -30 & 0 & 0 & 0 & 1 & | & 0 \end{bmatrix}$$

$$\begin{bmatrix} 0 & 10 & 2 & 1 & 0 & 2 & 0 & | & 300 \\ 0 & -43/3 & -19/3 & 0 & 1 & -5/3 & 0 & | & 380 \\ 1 & 1/3 & 1/3 & 0 & 0 & -1/3 & 0 & | & 100 \\ 0 & -140/3 & -80/3 & 0 & 0 & -10/3 & 1 & | & 1000 \end{bmatrix}$$

$$\begin{bmatrix} 0 & 1 & 1/5 & 1/10 & 0 & 1/5 & 0 & | & 30 \\ 0 & 0 & -52/15 & 43/30 & 1 & 6/5 & 0 & | & 810 \\ 1 & 0 & 4/15 & -1/30 & 0 & -2/5 & 0 & | & 90 \\ 0 & 0 & -52/3 & 14/3 & 0 & 6 & 1 & | & 2400 \end{bmatrix}$$

$$\begin{bmatrix} 0 & 5 & 1 & 1/2 & 0 & 1 & 0 & | & 150 \\ 0 & 52/3 & 0 & 19/6 & 1 & 14/3 & 0 & | & 1330 \\ 1 & -4/3 & 0 & -1/6 & 0 & -2/3 & 0 & | & 50 \\ 0 & 260/3 & 0 & 40/3 & 0 & 70/3 & 1 & | & 5000 \end{bmatrix}$$

Maximum $z = 5000$ at $(50, 0, 150)$

17. In order to set up the initial simplex tableau we replace the objective function with
maximize $w = -15x_1 - 8x_2$
and replace the second constraint with
$$-3x_1 - 2x_2 \leq -36.$$
This gives the initial tableau

$$\begin{bmatrix} 1 & 2 & 1 & 0 & 0 & | & 20 \\ -3 & -2 & 0 & 1 & 0 & | & -36 \\ 15 & 8 & 0 & 0 & 1 & | & 0 \end{bmatrix}$$

In order to obtain a tableau with a basic feasible solution we pivot on -3 to obtain

$$\begin{bmatrix} 0 & 4/3 & 1 & 1/3 & 0 & | & 8 \\ 1 & 2/3 & 0 & -1/3 & 0 & | & 12 \\ 0 & -2 & 0 & 5 & 1 & | & -180 \end{bmatrix}$$

Now the pivot column is column 2, and the pivot row is row 1. Thus, we pivot on 4/3 and obtain

$$\left[\begin{array}{ccccc|c} 0 & 1 & 3/4 & 1/4 & 0 & 6 \\ 1 & 0 & -1/2 & -1/2 & 0 & 8 \\ 0 & 0 & 3/2 & 11/2 & 1 & -168 \end{array}\right]$$

This gives the optimal solution maximum w = − 168 at (8, 6) so the original problem has the minimum value z = 168 at (8, 6).

19.
$$\left[\begin{array}{cccccc|c} -10 & -12 & -5 & 1 & 0 & 0 & -100 \\ 5 & 7 & 5 & 0 & 1 & 0 & 75 \\ 4 & 5 & 1 & 0 & 0 & 1 & 0 \end{array}\right]$$

$$\left[\begin{array}{cccccc|c} 5/6 & 1 & 5/12 & -1/12 & 0 & 0 & 25/3 \\ -5/6 & 0 & 25/12 & 7/12 & 1 & 0 & 50/3 \\ -1/6 & 0 & -13/12 & 5/12 & 0 & 1 & -125/3 \end{array}\right]$$

$$\left[\begin{array}{cccccc|c} 1 & 1 & 0 & -1/5 & -1/5 & 0 & 5 \\ -2/5 & 0 & 1 & 7/25 & 12/25 & 0 & 8 \\ -3/5 & 0 & 0 & 18/25 & 13/25 & 1 & -33 \end{array}\right]$$

$$\left[\begin{array}{cccccc|c} 1 & 1 & 0 & -1/5 & -1/5 & 0 & 5 \\ 0 & 2/5 & 1 & 1/5 & 2/5 & 0 & 10 \\ 0 & 3/5 & 0 & 3/5 & 2/5 & 1 & -30 \end{array}\right]$$ Minimum z = 30 at (5, 0, 10)

21.
$$\left[\begin{array}{ccccccc|c} 3 & 2 & 1 & 0 & 0 & 0 & 0 & 48 \\ 2 & 4 & 0 & 1 & 0 & 0 & 0 & 64 \\ -4 & -6 & 0 & 0 & 1 & 0 & 0 & -84 \\ -8 & -4 & 0 & 0 & 0 & 0 & 1 & 0 \end{array}\right]$$

$$\left[\begin{array}{ccccccc|c} 5/3 & 0 & 1 & 0 & 1/3 & 0 & 0 & 20 \\ -2/3 & 0 & 0 & 1 & 2/3 & 0 & 0 & 8 \\ 2/3 & 1 & 0 & 0 & -1/6 & 0 & 0 & 14 \\ -16/3 & 0 & 0 & 0 & -2/3 & 0 & 1 & 56 \end{array}\right]$$

$$\left[\begin{array}{cccccc|c} 1 & 0 & 3/5 & 0 & 1/5 & 0 & 12 \\ 0 & 0 & 2/5 & 1 & 4/5 & 0 & 16 \\ 0 & 1 & -2/5 & 0 & -3/10 & 0 & 6 \\ 0 & 0 & 16/5 & 0 & 2/5 & 1 & 120 \end{array}\right]$$ Maximum z = 120 at (12, 6)

23.
$$\left[\begin{array}{ccccccc|c} 3 & 2 & 1 & 0 & 0 & 0 & 0 & 60 \\ -2 & -3 & 0 & 1 & 0 & 0 & 0 & -24 \\ 1 & 1 & 0 & 0 & 1 & 0 & 0 & 25 \\ -1 & -1 & 0 & 0 & 0 & 1 & 0 & -25 \\ -6 & -4 & 0 & 0 & 0 & 0 & 1 & 0 \end{array}\right]$$

$$\left[\begin{array}{ccccccc|c} 0 & -1 & 1 & 0 & 0 & 3 & 0 & -15 \\ 0 & -1 & 0 & 1 & 0 & -2 & 0 & 26 \\ 0 & 0 & 0 & 0 & 1 & 1 & 0 & 0 \\ 1 & 1 & 0 & 0 & 0 & -1 & 0 & 25 \\ 0 & 2 & 0 & 0 & 0 & -6 & 1 & 150 \end{array}\right]$$

$$\left[\begin{array}{ccccccc|c}
0 & 1 & -1 & 0 & 0 & -3 & 0 & 15 \\
0 & 0 & -1 & 1 & 0 & -5 & 0 & 41 \\
0 & 0 & 0 & 0 & 1 & 1 & 0 & 0 \\
1 & 0 & 1 & 0 & 0 & 2 & 0 & 10 \\
\hline
0 & 0 & 2 & 0 & 0 & 0 & 1 & 120
\end{array}\right]$$

Maximum z = 120 at (10, 15)

25.
$$\left[\begin{array}{cccccccc|c}
7 & 12 & 12 & 1 & 0 & 0 & 0 & 0 & 312 \\
-13 & -20 & -12 & 0 & 1 & 0 & 0 & 0 & -384 \\
5 & 4 & 12 & 0 & 0 & 1 & 0 & 0 & 168 \\
-5 & -4 & -12 & 0 & 0 & 0 & 1 & 0 & -168 \\
\hline
-10 & -24 & -26 & 0 & 0 & 0 & 0 & 1 & 0
\end{array}\right]$$

$$\left[\begin{array}{cccccccc|c}
-4/5 & 0 & 24/5 & 1 & 3/5 & 0 & 0 & 0 & 408/5 \\
13/20 & 1 & 3/5 & 0 & -1/20 & 0 & 0 & 0 & 96/5 \\
12/5 & 0 & 48/5 & 0 & 1/5 & 1 & 0 & 0 & 456/5 \\
-12/5 & 0 & -48/5 & 0 & -1/5 & 0 & 1 & 0 & -456/5 \\
\hline
28/5 & 0 & -58/5 & 0 & -6/5 & 0 & 0 & 1 & 2304/5
\end{array}\right]$$

$$\left[\begin{array}{ccccccc|c}
-2 & 0 & 0 & 1 & 1/2 & 0 & 1/2 & 0 & 36 \\
1/2 & 1 & 0 & 0 & -1/16 & 0 & 1/16 & 0 & 27/2 \\
0 & 0 & 0 & 0 & 0 & 1 & 1 & 0 & 0 \\
1/4 & 0 & 1 & 0 & 1/48 & 0 & -5/48 & 0 & 19/2 \\
\hline
17/2 & 0 & 0 & 0 & -23/24 & 0 & -29/24 & 1 & 571
\end{array}\right]$$

$$\left[\begin{array}{ccccccc|c}
-2 & 0 & 0 & 1 & 1/2 & -1/2 & 0 & 0 & 36 \\
1/2 & 1 & 0 & 0 & -1/16 & -1/16 & 0 & 0 & 27/2 \\
0 & 0 & 0 & 0 & 0 & 1 & 1 & 0 & 0 \\
1/4 & 0 & 1 & 0 & 1/48 & 5/48 & 0 & 0 & 19/2 \\
\hline
17/2 & 0 & 0 & 0 & -23/24 & 29/24 & 0 & 1 & 571
\end{array}\right]$$

$$\left[\begin{array}{ccccccc|c}
-4 & 0 & 0 & 2 & 1 & -1 & 0 & 0 & 72 \\
1/4 & 1 & 0 & 1/8 & 0 & -1/8 & 0 & 0 & 18 \\
0 & 0 & 0 & 0 & 0 & 1 & 1 & 0 & 0 \\
1/3 & 0 & 1 & -1/24 & 0 & 1/8 & 0 & 0 & 8 \\
\hline
14/3 & 0 & 0 & 23/12 & 0 & 1/4 & 0 & 1 & 640
\end{array}\right]$$

Maximum z = 640 at (0,18,8)

27.
$$\left[\begin{array}{ccccccc|c}
-2 & 5 & 1 & 0 & 0 & 0 & 0 & 90 \\
4 & 3 & 0 & 1 & 0 & 0 & 0 & 80 \\
-4 & -3 & 0 & 0 & 1 & 0 & 0 & -80 \\
-2 & 1 & 0 & 0 & 0 & 1 & 0 & -20 \\
\hline
9 & 5 & 0 & 0 & 0 & 0 & 1 & 0
\end{array}\right]$$

$$\left[\begin{array}{ccccccc|c}
0 & 13/2 & 1 & 0 & -1/2 & 0 & 0 & 130 \\
0 & 0 & 0 & 1 & 1 & 0 & 0 & 0 \\
1 & 3/4 & 0 & 0 & -1/4 & 0 & 0 & 20 \\
0 & 5/2 & 0 & 0 & -1/2 & 1 & 0 & 20 \\
\hline
0 & -7/4 & 0 & 0 & 9/4 & 0 & 1 & -180
\end{array}\right]$$

$$\left[\begin{array}{ccccccc|c}
0 & 0 & 1 & 0 & 4/5 & -13/5 & 0 & 78 \\
0 & 0 & 0 & 1 & 1 & 0 & 0 & 0 \\
1 & 0 & 0 & 0 & -1/10 & -3/10 & 0 & 14 \\
0 & 1 & 0 & 0 & -1/5 & 2/5 & 0 & 8 \\
\hline
0 & 0 & 0 & 0 & 19/10 & 7/10 & 1 & -166
\end{array}\right]$$

Minimum z is 166 at $x_1 = 14$, $x_2 = 8$

89

29.

$$
\left[\begin{array}{cccccccc|c}
-10 & -12 & -5 & 1 & 0 & 0 & 0 & 0 & -100 \\
5 & 7 & 5 & 0 & 1 & 0 & 0 & 0 & 75 \\
-10 & -2 & -10 & 0 & 0 & 1 & 0 & 0 & -120 \\
\underline{10} & \underline{2} & \underline{10} & \underline{0} & \underline{0} & \underline{0} & \underline{1} & \underline{0} & \underline{120} \\
8 & 10 & 2 & 0 & 0 & 0 & 0 & 1 & 0
\end{array}\right]
$$

$$
\left[\begin{array}{cccccccc|c}
-5 & -11 & 0 & 1 & 0 & -1/2 & 0 & 0 & -40 \\
0 & 6 & 0 & 0 & 1 & 1/2 & 0 & 0 & 15 \\
1 & 1/5 & 1 & 0 & 0 & -1/10 & 0 & 0 & 12 \\
\underline{0} & \underline{0} & \underline{0} & \underline{0} & \underline{0} & \underline{1} & \underline{1} & \underline{0} & \underline{0} \\
6 & 48/5 & 0 & 0 & 0 & 1/5 & 0 & 1 & -24
\end{array}\right]
$$

$$
\left[\begin{array}{cccccccc|c}
1 & 11/5 & 0 & -1/5 & 0 & 1/10 & 0 & 0 & 8 \\
0 & 6 & 0 & 0 & 1 & 1/2 & 0 & 0 & 15 \\
0 & -2 & 1 & 1/5 & 0 & -1/5 & 0 & 0 & 4 \\
\underline{0} & \underline{0} & \underline{0} & \underline{0} & \underline{0} & \underline{1} & \underline{1} & \underline{0} & \underline{0} \\
0 & -18/5 & 0 & 6/5 & 0 & -2/5 & 0 & 1 & -72
\end{array}\right]
$$

$$
\left[\begin{array}{cccccccc|c}
1 & 0 & 0 & -1/5 & -11/30 & -1/12 & 0 & 0 & 5/2 \\
0 & 1 & 0 & 0 & 1/6 & 1/12 & 0 & 0 & 5/2 \\
0 & 0 & 1 & 1/5 & 1/3 & -1/30 & 0 & 0 & 9 \\
\underline{0} & \underline{0} & \underline{0} & \underline{0} & \underline{0} & \underline{1} & \underline{1} & \underline{0} & \underline{0} \\
0 & 0 & 0 & 6/5 & 3/5 & -1/10 & 0 & 1 & -63
\end{array}\right]
$$

$$
\left[\begin{array}{cccccccc|c}
1 & 0 & 0 & -1/5 & -11/30 & 0 & 1/12 & 0 & 5/2 \\
0 & 1 & 0 & 0 & 1/6 & 0 & -1/12 & 0 & 5/2 \\
0 & 0 & 1 & 1/5 & 1/3 & 0 & 1/30 & 0 & 9 \\
\underline{0} & \underline{0} & \underline{0} & \underline{0} & \underline{0} & \underline{1} & \underline{1} & \underline{0} & \underline{0} \\
0 & 0 & 0 & 6/5 & 3/5 & 0 & 1/10 & 1 & -63
\end{array}\right]
$$

Minimum $z = 63$ at $x_1 = 5/2$, $x_2 = 5/2$, $x_3 = 9$

31. Let x = number of Custom, y = number of Executive.
Minimize $C = 70x + 80y$
Subject to
$$x + y \geq 100$$
$$90x + 120y \geq 10800$$
$$4x + 5y \leq 800$$
$$x \geq 0, y \geq 0$$

$$
\left[\begin{array}{cccccc|c}
-1 & -1 & 1 & 0 & 0 & 0 & -100 \\
-90 & -120 & 0 & 1 & 0 & 0 & -10800 \\
\underline{4} & \underline{5} & \underline{0} & \underline{0} & \underline{1} & \underline{0} & \underline{800} \\
70 & 80 & 0 & 0 & 0 & 1 & 0
\end{array}\right]
$$

33. Let x = number of A, y = number of B, z = number of C.
Maximize $p = 6x + 9y + 6z$
Subject to
$$x + y + z \geq 6600$$
$$20x + 25y + 15z \leq 133000$$
$$x + 3y + 2z \leq 13600$$
$$8x + 10y + 15z \leq 73000$$
$$x \geq 0, y \geq 0, z \geq 0$$

$$\begin{bmatrix} -1 & -1 & -1 & 1 & 0 & 0 & 0 & 0 & -6600 \\ 20 & 25 & 15 & 0 & 1 & 0 & 0 & 0 & 133000 \\ 1 & 3 & 2 & 0 & 0 & 1 & 0 & 0 & 13600 \\ 8 & 10 & 15 & 0 & 0 & 0 & 1 & 0 & 73000 \\ -6 & -9 & -6 & 0 & 0 & 0 & 0 & 1 & 0 \end{bmatrix}$$

35.
$$\begin{bmatrix} -1 & -1 & -1 & 1 & 0 & 0 & 0 & 0 & 0 & -6800 \\ 20 & 25 & 15 & 0 & 1 & 0 & 0 & 0 & 0 & 133000 \\ 1 & 3 & 2 & 0 & 0 & 1 & 0 & 0 & 0 & 13600 \\ 8 & 10 & 15 & 0 & 0 & 0 & 1 & 0 & 0 & 80000 \\ 0 & -1 & 0 & 0 & 0 & 0 & 0 & 1 & 0 & -2000 \\ -6 & -9 & -6 & 0 & 0 & 0 & 0 & 0 & 1 & 0 \end{bmatrix}$$

37. Let x_1 = minutes jogging, x_2 = minutes playing handball, x_3 = minutes swimming.

(a) Minimize $z = x_1 + x_2 + x_3$

Subject to $13x_1 + 11x_2 + 7x_3 \geq 660$

$x_1 - x_3 = 0$

$x_2 \geq 2x_1$

$x_1 \geq 0, x_2 \geq 0, x_3 \geq 0$

$$\begin{bmatrix} -13 & -11 & -7 & 1 & 0 & 0 & 0 & 0 & -660 \\ 1 & 0 & -1 & 0 & 1 & 0 & 0 & 0 & 0 \\ -1 & 0 & 1 & 0 & 0 & 1 & 0 & 0 & 0 \\ 2 & -1 & 0 & 0 & 0 & 0 & 1 & 0 & 0 \\ 1 & 1 & 1 & 0 & 0 & 0 & 0 & 1 & 0 \end{bmatrix}$$

(b) Maximize $z = 13x_1 + 11x_2 + 7x_3$

Subject to $x_1 + x_2 + x_3 \leq 90$

$x_1 - x_3 = 0$

$x_2 \geq 2x_1$

$x_1 \geq 0, x_2 \geq 0, x_3 \geq 0$

$$\begin{bmatrix} 1 & 1 & 1 & 1 & 0 & 0 & 0 & 0 & 90 \\ 1 & 0 & -1 & 0 & 1 & 0 & 0 & 0 & 0 \\ -1 & 0 & 1 & 0 & 0 & 1 & 0 & 0 & 0 \\ 2 & -1 & 0 & 0 & 0 & 0 & 1 & 0 & 0 \\ -13 & -11 & -7 & 0 & 0 & 0 & 0 & 1 & 0 \end{bmatrix}$$

39. Let x_1 = number pounds of Lite, x_2 = number pounds of Trim, x_3 = number pounds of Regular, x_4 = number pounds of Health Fare
Maximize profit

$z = .25x_1 + .25x_2 + .27x_3 + .32x_4$ subject to

$.75x_1 + .50x_2 + .80x_3 + .15x_4 \leq 2400$ (Wheat)

$.20x_1 + .25x_2 + .20x_3 + .50x_4 \leq 1400$ (Oats)

$.05x_1 + .20x_2 \qquad + .25x_4 \leq 700$ (Raisins)

$.05x_2 \qquad + .10x_4 \leq 250$ (Nuts, maximum)

$.05x_2 \qquad + .10x_4 \geq 200$ (Nuts, minimum)

$x_1 \geq 0, x_2 \geq 0, x_2 \geq 0, x_4 \geq 0$

Initial Tableau

$$
\begin{bmatrix}
.75 & .50 & .80 & .15 & 1 & 0 & 0 & 0 & 0 & 0 & 2400 \\
.20 & .25 & .20 & .50 & 0 & 1 & 0 & 0 & 0 & 0 & 1400 \\
.05 & .20 & 0 & .25 & 0 & 0 & 1 & 0 & 0 & 0 & 700 \\
0 & .05 & 0 & .10 & 0 & 0 & 0 & 1 & 0 & 0 & 250 \\
0 & -.05 & 0 & -.10 & 0 & 0 & 0 & 0 & 1 & 0 & -200 \\
\hline
-.25 & -.25 & -.27 & -.32 & 0 & 0 & 0 & 0 & 0 & 1 & 0
\end{bmatrix}
$$

41. Let x_1 = number units of food A, x_2 = number units of food B,
    x_3 = number units of food C

Minimize cost
$$z = 1.40x_1 + 1.65x_2 + 1.95x_3 \quad \text{subject to}$$
$$15x_1 + 10x_2 + 23x_3 \geq 80 \quad \text{(protein)}$$
$$20x_1 + 30x_2 + 11x_3 \geq 95 \quad \text{(carbohydrates)}$$
$$500x_1 + 400x_2 + 200x_3 \geq 3800 \quad \text{(calories)}$$
$$8x_1 + 3x_2 + 6x_3 \leq 35 \quad \text{(fat)}$$
$$x_1 \geq 0, \ x_2 \geq 0, \ x_3 \geq 0$$

Initial Tableau

$$
\begin{bmatrix}
-15 & -10 & -23 & 1 & 0 & 0 & 0 & 0 & -80 \\
-20 & -30 & -11 & 0 & 1 & 0 & 0 & 0 & -95 \\
-500 & -400 & -200 & 0 & 0 & 1 & 0 & 0 & -3800 \\
8 & 3 & 6 & 0 & 0 & 0 & 1 & 0 & 35 \\
\hline
1.40 & 1.65 & 1.95 & 0 & 0 & 0 & 0 & 1 & 0
\end{bmatrix}
$$

43. Let x = number of Custom, y = number of Executive.
    Minimize $C = 70x + 80y$
    Subject to
$$x + y \geq 100$$
$$90x + 120y \geq 10800$$
$$4x + 5y \leq 800$$

$$
\begin{bmatrix}
-1 & -1 & 1 & 0 & 0 & 0 & -100 \\
-90 & -120 & 0 & 1 & 0 & 0 & -10800 \\
4 & 5 & 0 & 0 & 1 & 0 & 800 \\
\hline
70 & 80 & 0 & 0 & 0 & 1 & 0
\end{bmatrix}
$$

$$
\begin{bmatrix}
-1/4 & 0 & 1 & -1/120 & 0 & 0 & -10 \\
3/4 & 1 & 0 & -1/120 & 0 & 0 & 90 \\
1/4 & 0 & 0 & 1/24 & 1 & 0 & 350 \\
\hline
10 & 0 & 0 & 2/3 & 0 & 1 & -7200
\end{bmatrix}
$$

$$
\begin{bmatrix}
1 & 0 & -4 & 1/30 & 0 & 0 & 40 \\
0 & 1 & 3 & -1/30 & 0 & 0 & 60 \\
0 & 0 & 1 & 1/30 & 1 & 0 & 340 \\
\hline
0 & 0 & 40 & 1/3 & 0 & 1 & -7600
\end{bmatrix}
$$

The manager should order 40 Custom and 60 Executive for a minimum cost of $7600.

45. Let x = number of A, y = number of B, z = number of C.
Maximize p = 6x + 9y + 6z
Subject to   x + y + z ≥ 6600
20x + 25y + 15z ≤ 133000
x + 3y + 2z ≤ 13600
8x + 10y + 15z ≤ 73000
x ≥ 0, y ≥ 0, z ≥ 0

$$\begin{bmatrix} -1 & -1 & -1 & 1 & 0 & 0 & 0 & 0 & -6600 \\ 20 & 25 & 15 & 0 & 1 & 0 & 0 & 0 & 133000 \\ 1 & 3 & 2 & 0 & 0 & 1 & 0 & 0 & 13600 \\ 8 & 10 & 15 & 0 & 0 & 0 & 1 & 0 & 73000 \\ -6 & -9 & -6 & 0 & 0 & 0 & 0 & 1 & 0 \end{bmatrix}$$

$$\begin{bmatrix} 1 & 1 & 1 & -1 & 0 & 0 & 0 & 0 & 6600 \\ -5 & 0 & -10 & 25 & 1 & 0 & 0 & 0 & -32000 \\ -2 & 0 & -1 & 3 & 0 & 1 & 0 & 0 & -6200 \\ -2 & 0 & 5 & 10 & 0 & 0 & 1 & 0 & 7000 \\ 3 & 0 & 3 & -9 & 0 & 0 & 0 & 1 & 59400 \end{bmatrix}$$

$$\begin{bmatrix} 0 & 1 & 1/2 & 1/2 & 0 & 1/2 & 0 & 0 & 3500 \\ 0 & 0 & -15/2 & 35/2 & 1 & -5/2 & 0 & 0 & -16500 \\ 1 & 0 & 1/2 & -3/2 & 0 & -1/2 & 0 & 0 & 3100 \\ 0 & 0 & 6 & 7 & 0 & -1 & 1 & 0 & 13200 \\ 0 & 0 & 3/2 & -9/2 & 0 & 3/2 & 0 & 1 & 50100 \end{bmatrix}$$

$$\begin{bmatrix} 0 & 1 & 0 & 5/3 & 1/15 & 1/3 & 0 & 0 & 2400 \\ 0 & 0 & 1 & -7/3 & -2/15 & 1/3 & 0 & 0 & 2200 \\ 1 & 0 & 0 & -1/3 & 1/15 & -2/3 & 0 & 0 & 2000 \\ 0 & 0 & 0 & 21 & 4/5 & -3 & 1 & 0 & 0 \\ 0 & 0 & 0 & -1 & 1/5 & 1 & 0 & 1 & 46800 \end{bmatrix}$$

$$\begin{bmatrix} 0 & 1 & 0 & 0 & 1/315 & 4/7 & -5/63 & 0 & 2400 \\ 0 & 0 & 1 & 0 & -2/45 & 0 & 1/9 & 0 & 2200 \\ 1 & 0 & 0 & 0 & 5/63 & -5/7 & 1/63 & 0 & 2000 \\ 0 & 0 & 0 & 1 & 4/105 & -1/7 & 1/21 & 0 & 0 \\ 0 & 0 & 0 & 0 & 5/21 & 6/7 & 1/21 & 1 & 46800 \end{bmatrix}$$

Maximum profit is $46,800 when 2000 of A, 2400 of B, and 2200 of C are ordered.

$$\begin{bmatrix} -1 & -1 & -1 & 1 & 0 & 0 & 0 & 0 & 0 & -6800 \\ 20 & 25 & 15 & 0 & 1 & 0 & 0 & 0 & 0 & 133000 \\ 1 & 3 & 2 & 0 & 0 & 1 & 0 & 0 & 0 & 13600 \\ 8 & 10 & 15 & 0 & 0 & 0 & 1 & 0 & 0 & 80000 \\ 0 & -1 & 0 & 0 & 0 & 0 & 0 & 1 & 0 & -2000 \\ -6 & -9 & -6 & 0 & 0 & 0 & 0 & 0 & 1 & 0 \end{bmatrix}$$

47.

$$\begin{bmatrix} 1 & 1 & 1 & -1 & 0 & 0 & 0 & 0 & 0 & 6800 \\ -5 & 0 & -10 & 25 & 1 & 0 & 0 & 0 & 0 & -37000 \\ -2 & 0 & -1 & 3 & 0 & 1 & 0 & 0 & 0 & -6800 \\ -2 & 0 & 5 & 10 & 0 & 0 & 1 & 0 & 0 & 12000 \\ 1 & 0 & 1 & -1 & 0 & 0 & 0 & 1 & 0 & 4800 \\ 3 & 0 & 3 & -9 & 0 & 0 & 0 & 0 & 1 & 61200 \end{bmatrix}$$

93

$$\begin{bmatrix} 0 & 1 & 1/2 & 1/2 & 0 & 1/2 & 0 & 0 & 0 & 3400 \\ 0 & 0 & -15/2 & 35/2 & 1 & -5/2 & 0 & 0 & 0 & -20000 \\ 1 & 0 & 1/2 & -3/2 & 0 & -1/2 & 0 & 0 & 0 & 3400 \\ 0 & 0 & 6 & 7 & 0 & -1 & 1 & 0 & 0 & 18800 \\ 0 & 0 & 1/2 & 1/2 & 0 & 1/2 & 0 & 1 & 0 & 1400 \\ \hline 0 & 0 & 3/2 & -9/2 & 0 & 3/2 & 0 & 0 & 1 & 51000 \end{bmatrix}$$

$$\begin{bmatrix} 0 & 1 & 0 & 5/3 & 1/15 & 1/3 & 0 & 0 & 0 & 6200/3 \\ 0 & 0 & 1 & -7/3 & -2/15 & 1/3 & 0 & 0 & 0 & 8000/3 \\ 1 & 0 & 0 & -1/3 & 1/15 & -2/3 & 0 & 0 & 0 & 6200/3 \\ 0 & 0 & 0 & 21 & 4/5 & -3 & 1 & 0 & 0 & 2800 \\ 0 & 0 & 0 & 5/3 & 1/15 & 1/3 & 0 & 1 & 0 & 200/3 \\ \hline 0 & 0 & 0 & -1 & 1/5 & 1 & 0 & 0 & 1 & 47000 \end{bmatrix}$$

$$\begin{bmatrix} 0 & 1 & 0 & 0 & 0 & 0 & 0 & -1 & 0 & 2000 \\ 0 & 0 & 1 & 0 & -1/25 & 4/5 & 0 & 7/5 & 0 & 2760 \\ 1 & 0 & 0 & 0 & 2/25 & -3/5 & 0 & 1/5 & 0 & 2080 \\ 0 & 0 & 0 & 0 & -1/25 & -36/5 & 1 & -63/5 & 0 & 1960 \\ 0 & 0 & 0 & 1 & 1/25 & 1/5 & 0 & 3/5 & 0 & 40 \\ \hline 0 & 0 & 0 & 0 & 6/25 & 6/5 & 0 & 3/5 & 1 & 47040 \end{bmatrix}$$

Maximum profit is \$47,040 when ordering 2080 of A, 2000 of B, and 2760 of C.

49. Let x_1 = minutes jogging, x_2 = minutes playing handball, x_3 = minutes swimming.

    (a) Minimize $z = x_1 + x_2 + x_3$
    Subject to $13x_1 + 11x_2 + 7x_3 \geq 660$

$$x_1 - x_3 = 0$$
$$x_2 \geq 2x_1$$
$$x_1 \geq 0, \ x_2 \geq 0, \ x_3 \geq 0$$

$$\begin{bmatrix} -13 & -11 & -7 & 1 & 0 & 0 & 0 & 0 & -660 \\ 1 & 0 & -1 & 0 & 1 & 0 & 0 & 0 & 0 \\ -1 & 0 & 1 & 0 & 0 & 1 & 0 & 0 & 0 \\ 2 & -1 & 0 & 0 & 0 & 0 & 1 & 0 & 0 \\ \hline 1 & 1 & 1 & 0 & 0 & 0 & 0 & 1 & 0 \end{bmatrix}$$

$$\begin{bmatrix} 13/11 & 1 & 7/11 & -1/11 & 0 & 0 & 0 & 0 & 60 \\ 1 & 0 & -1 & 0 & 1 & 0 & 0 & 0 & 0 \\ -1 & 0 & 1 & 0 & 0 & 1 & 0 & 0 & 0 \\ 35/11 & 0 & 7/11 & -1/11 & 0 & 0 & 1 & 0 & 60 \\ \hline -2/11 & 0 & 4/11 & 1/11 & 0 & 0 & 0 & 1 & -60 \end{bmatrix}$$

$$\begin{bmatrix} 0 & 1 & 20/11 & -1/11 & -13/11 & 0 & 0 & 0 & 60 \\ 1 & 0 & -1 & 0 & 1 & 0 & 0 & 0 & 0 \\ 0 & 0 & 0 & 0 & 1 & 1 & 0 & 0 & 0 \\ 0 & 0 & 42/11 & -1/11 & -35/11 & 0 & 1 & 0 & 60 \\ \hline 0 & 0 & 2/11 & 1/11 & 2/11 & 0 & 0 & 1 & -60 \end{bmatrix}$$

0 minutes jogging, 60 minutes playing handball, 0 minutes swimming.

(b)   Maximize $z = 13x_1 + 11x_2 + 7x_3$
Subject to $x_1 + x_2 + x_3 \leq 90$
$$x_1 - x_3 = 0$$
$$x_2 \geq 2x_1$$
$$x_1 \geq 0, \ x_2 \geq 0, \ x_3 \geq 0$$

$$\left[\begin{array}{cccccccc|c}
1 & 1 & 1 & 1 & 0 & 0 & 0 & 0 & 90 \\
1 & 0 & -1 & 0 & 1 & 0 & 0 & 0 & 0 \\
-1 & 0 & 1 & 0 & 0 & 1 & 0 & 0 & 0 \\
2 & -1 & 0 & 0 & 0 & 0 & 1 & 0 & 0 \\
-13 & -11 & -7 & 0 & 0 & 0 & 0 & 1 & 0
\end{array}\right]$$

$$\left[\begin{array}{cccccccc|c}
0 & 1 & 2 & 1 & -1 & 0 & 0 & 0 & 90 \\
1 & 0 & -1 & 0 & 1 & 0 & 0 & 0 & 0 \\
0 & 0 & 0 & 0 & 1 & 1 & 0 & 0 & 0 \\
0 & -1 & 2 & 0 & -2 & 0 & 1 & 0 & 0 \\
0 & -11 & -20 & 0 & 13 & 0 & 0 & 1 & 0
\end{array}\right]$$

$$\left[\begin{array}{cccccccc|c}
0 & 2 & 0 & 1 & 1 & 0 & -1 & 0 & 90 \\
1 & -1/2 & 0 & 0 & 0 & 0 & 1/2 & 0 & 0 \\
0 & 0 & 0 & 0 & 1 & 1 & 0 & 0 & 0 \\
0 & -1/2 & 1 & 0 & -1 & 0 & 1/2 & 0 & 0 \\
0 & -21 & 0 & 0 & -7 & 0 & 10 & 1 & 0
\end{array}\right]$$

$$\left[\begin{array}{cccccccc|c}
0 & 1 & 0 & 1/2 & 1/2 & 0 & -1/2 & 0 & 45 \\
1 & 0 & 0 & 1/4 & 1/4 & 0 & 1/4 & 0 & 45/2 \\
0 & 0 & 0 & 0 & 1 & 1 & 0 & 0 & 0 \\
0 & 0 & 1 & 1/4 & -3/4 & 0 & 1/4 & 0 & 45/2 \\
0 & 0 & 0 & 21/2 & 7/2 & 0 & -1/2 & 1 & 945
\end{array}\right]$$

$$\left[\begin{array}{cccccccc|c}
2 & 1 & 0 & 1 & 1 & 0 & 0 & 0 & 90 \\
4 & 0 & 0 & 1 & 1 & 0 & 1 & 0 & 90 \\
0 & 0 & 0 & 0 & 1 & 1 & 0 & 0 & 0 \\
-1 & 0 & 1 & 0 & -1 & 0 & 0 & 0 & 0 \\
2 & 0 & 0 & 11 & 4 & 0 & 0 & 1 & 990
\end{array}\right]$$

No time jogging or swimming - all 90 minutes playing handball

51.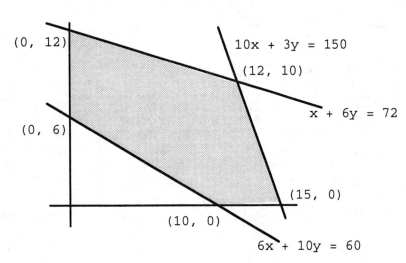

Initial Tableau

$$\begin{bmatrix} -6 & -10 & 1 & 0 & 0 & 0 & -60 \\ 1 & 6 & 0 & 1 & 0 & 0 & 72 \\ \underline{10} & \underline{3} & \underline{0} & \underline{0} & \underline{1} & \underline{0} & \underline{150} \\ -8 & -10 & 0 & 0 & 0 & 1 & 0 \end{bmatrix}$$

Corner (0, 0), z = 0 Not a feasible solution

Tableau 1

$$\begin{bmatrix} 3/5 & 1 & -1/10 & 0 & 0 & 0 & 6 \\ -13/5 & 0 & 3/5 & 1 & 0 & 0 & 36 \\ \underline{41/5} & \underline{0} & \underline{3/10} & \underline{0} & \underline{1} & \underline{0} & \underline{132} \\ -2 & 0 & -1 & 0 & 0 & 1 & 60 \end{bmatrix}$$

Corner (0, 6), z = 60 Feasible solution

Tableau 2

$$\begin{bmatrix} 1 & 5/3 & -1/6 & 0 & 0 & 0 & 10 \\ 0 & 13/3 & 1/6 & 1 & 0 & 0 & 62 \\ \underline{0} & \underline{-41/3} & \underline{5/3} & \underline{0} & \underline{1} & \underline{0} & \underline{50} \\ 0 & 10/3 & -4/3 & 0 & 0 & 1 & 80 \end{bmatrix}$$

Corner (10, 0), z = 80 Feasible solution

Tableau 3

$$\begin{bmatrix} 1 & 3/10 & 0 & 0 & 1/10 & 0 & 15 \\ 0 & 57/10 & 0 & 1 & -1/10 & 0 & 57 \\ \underline{0} & \underline{-41/5} & \underline{1} & \underline{0} & \underline{3/5} & \underline{0} & \underline{30} \\ 0 & -38/5 & 0 & 0 & 4/5 & 1 & 120 \end{bmatrix}$$

Corner (15, 0), z = 120 Feasible solution

Tableau 4

$$\begin{bmatrix} 1 & 0 & 0 & -1/19 & 2/19 & 0 & 12 \\ 0 & 1 & 0 & 10/57 & -1/57 & 0 & 10 \\ \underline{0} & \underline{0} & \underline{1} & \underline{82/57} & \underline{26/57} & \underline{0} & \underline{112} \\ 0 & 0 & 0 & 4/3 & 2/3 & 1 & 196 \end{bmatrix}$$

Corner (12, 10), z = 196 Feasible solution

Corner	z	Feasible?
(0, 0)	0	no
(0, 6)	60	yes
(10, 0)	80	yes
(15, 0)	120	yes
(12, 10)	196	yes

The simplex method starts at (0,0), which is not feasible, then moves to (0,6), which is feasible. From there on each pivot moves to an adjacent corner, which is in the feasible region.

If the first pivot had been on -6 in row 1, column 1, the sequence of corners would be (0, 0), (10, 0), (15, 0), (12, 0).

53.   (a)

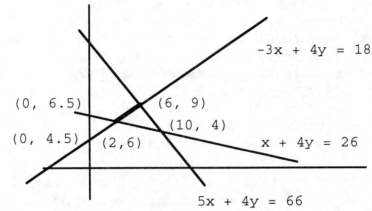

(0, 6.5)   (6, 9)
              (10, 4)
(0, 4.5) (2,6)   x + 4y = 26

            5x + 4y = 66

-3x + 4y = 18

(b)   The feasible region is the portion on the line $-3x + 4y = 18$
       that lies in the region bounded by $5x + 4y \leq 66$, by $x + 4y \geq$
       26. This is simply the line segment joining (2, 6) and (6, 9).
(c)   The feasible region is the portion of the line
       $a_1x + b_1y = k_1$ that lies in the region bounded by
       $a_2x + b_2y \leq k_2$ and
       $a_3x + b_3y \geq k_3$

55.   Let the variables be the following:
           x_1, x_2, and x_3 = quantity shipped from Cleveland to Chicago,
       Dallas, and Atlanta, respectively.
           x_4, x_5, and x_6 = quantity shipped from St. Louis to Chicago,
       Dallas, and Atlanta, respectively.
           x_7, x_8, and x_9 = quantity shipped from Pittsburgh to
       Chicago, Dallas, and Atlanta, respectively.
       (a)   The constraints are
           $x_1 + x_2 + x_3 \leq 4200$ (Quantity shipped from Cleveland
           does not exceed production)
           $x_4 + x_5 + x_6 \leq 4800$ (Quantity shipped from St. Louis)
           $x_7 + x_8 + x_9 \leq 3700$ (Quantity shipped from Pittsburgh)
           $x_1 + x_4 + x_7 = 5400$ (Quantity shipped to Chicago)
           $x_2 + x_5 + x_8 = 3800$ (Quantity shipped to Dallas)
           $x_3 + x_6 + x_9 = 2700$ (Quantity shipped to Atlanta)
           with all variables nonnegative.
       Since we are to minimize shipping costs, the objective function is
           Minimize $z = 35x_1 + 64x_2 + 60x_3 + 37x_4 + 59x_5 + 51x_6 +$
               $49x_7 + 68x_8 + 57x_9$
       (b)   The initial simplex tableau is

x_1	x_2	x_3	x_4	x_5	x_6	x_7	x_8	x_9	s_1	s_2	s_3	s_4	s_5	s_6	s_7	s_8	s_9	z	
1	1	1	0	0	0	0	0	0	1	0	0	0	0	0	0	0	0	0	4200
0	0	0	1	1	1	0	0	0	0	1	0	0	0	0	0	0	0	0	4800
0	0	0	0	0	0	1	1	1	0	0	1	0	0	0	0	0	0	0	3700
1	0	0	1	0	0	1	0	0	0	0	0	1	0	0	0	0	0	0	5400
-1	0	0	-1	0	0	-1	0	0	0	0	0	0	1	0	0	0	0	0	-5400
0	1	0	0	1	0	0	1	0	0	0	0	0	0	1	0	0	0	0	3800
0	-1	0	0	-1	0	0	-1	0	0	0	0	0	0	0	1	0	0	0	-3800
0	0	1	0	0	1	0	0	1	0	0	0	0	0	0	0	1	0	0	2700
0	0	-1	0	0	-1	0	0	-1	0	0	0	0	0	0	0	0	1	0	-2700
35	64	60	37	59	51	49	68	57	0	0	0	0	0	0	0	0	0	1	0

97

57. (a) x_1, x_2, x_3 represent the assignment of Arnold to Acme, Johnson, or Family, respectively.
x_4, x_5, x_6 represent the assignment of Bea to Acme, Johnson, or Family, respectively.
x_7, x_8, x_9 represent the assignment of Carl to Acme, Johnson, or Family, respectively.
x_{10}, x_{11}, x_{12} represent the assignment of Dee to Acme, Johnson, or Family, respectively.
$x_1 + x_2 + x_3 \le 1$ (Arnold assigned to at most one client.)
$x_4 + x_5 + x_6 \le 1$ (Bea assigned to at most one client.)
$x_7 + x_8 + x_9 \le 1$ (Carl assigned to at most one client.)
$x_{10} + x_{11} + x_{12} \le 1$ (Dee issigned to at most one client.)
$x_1 + x_4 + x_7 + x_{10} = 1$ (One auditor is assigned Acme.)
$x_2 + x_5 + x_8 + x_{11} = 1$ (One auditor is assigned Johnson.)
$x_3 + x_6 + x_9 + x_{12} = 1$ (One auditor is assigned Family.)
All variables are nonnegative.
Minimize total time: $z = 10x_1 + 15x_2 + 9x_3 + 9x_4 + 18x_5 + 5x_6 + 6x_7 + 14x_8 + 3x_9 + 8x_{10} + 16x_{11} + 6x_{12}$

(b)

```
 1  1  1  0  0  0  0  0  0  0  0  0  1  0  0  0  0  0  0  0  0  0 |  1
 0  0  0  1  1  1  0  0  0  0  0  0  0  1  0  0  0  0  0  0  0  0 |  1
 0  0  0  0  0  0  1  1  1  0  0  0  0  0  1  0  0  0  0  0  0  0 |  1
 0  0  0  0  0  0  0  0  0  1  1  1  0  0  0  1  0  0  0  0  0  0 |  1
 1  0  0  1  0  0  1  0  0  1  0  0  0  0  0  0  1  0  0  0  0  0 |  1
-1  0  0 -1  0  0 -1  0  0 -1  0  0  0  0  0  0  0  1  0  0  0  0 | -1
 0  1  0  0  1  0  0  1  0  0  1  0  0  0  0  0  0  0  1  0  0  0 |  1
 0 -1  0  0 -1  0  0 -1  0  0 -1  0  0  0  0  0  0  0  0  1  0  0 | -1
 0  0  1  0  0  1  0  0  1  0  0  1  0  0  0  0  0  0  0  0  1  0 |  1
 0  0 -1  0  0 -1  0  0 -1  0  0 -1  0  0  0  0  0  0  0  0  0  1 | -1
10 15  9  9 18  5  6 14  3  8 16  6  0  0  0  0  0  0  0  0  0  0  1 | 0
```

(c) The optimal solution is not unique. A minimum of 26 hours is required and may be obtained by assigning Arnold to Johnson Roofing, Bea to Family Bookstore, and Carl to Acme. Also obtained by assigning Arnold to Johnson Roofing, Carl to Family Bookstore and Dee to Acme.

59.

```
  5    1    1    8   1   0   0   0   0 |   90
  2    4    3    2   0   1   0   0   0 |   81
 -2   -1   -1   -4   0   0   1   0   0 |  -45
-45  -27  -18  -36   0   0   0   0   1 |    0
```

```
  1   -1   -1   0   1   0    2   0 |    0
  1  3.5  2.5   0   0   1   .5   0 | 58.5
 .5  .25  .25   1   0   0 -.25   0 | 11.25
-27  -18   -9   0   0   0   -9   1 |  405
```

```
 1   -1   -1   0    1   0    2   0 |    0
 0  4.5  3.5   0   -1   1 -1.5   0 | 58.6
 0  .75  .75   1  -.5   0 -1.25  0 | 11.25
 0  -45  -36   0   27   0   45   1 |  405
```

98

$$\begin{bmatrix} 1 & 0 & -.22 & 0 & .78 & .22 & 1.67 & 0 & 13 \\ 0 & 1 & .78 & 0 & -.22 & .22 & -.33 & 0 & 13 \\ 0 & 0 & .17 & 1 & -.33 & -.17 & -1 & 0 & 1.5 \\ 0 & 0 & -1 & 0 & 17 & 10 & 30 & 1 & 990 \end{bmatrix}$$

$$\begin{bmatrix} 1 & 0 & 0 & 1.33 & .33 & 0 & .33 & 0 & 15 \\ 0 & 1 & 0 & -4.67 & 1.33 & 1 & 4.33 & 0 & 6 \\ 0 & 0 & 1 & 6.00 & -2.00 & -1 & -6.00 & 0 & 9 \\ 0 & 0 & 0 & 6 & 15 & 9 & 24 & 1 & 999 \end{bmatrix}$$

The maximum value of z is 999 and it occurs at (15, 6, 9, 0).

Section 4.5

1. The -20 in the last column with nonnegative entries in the rest of the first row indicates no feasible solutions.

3. There is an unbounded feasible region. Therefore there is no solution, because all entries in column 4 are negative.

5.
$$\begin{bmatrix} 3 & 5 & 1 & 0 & 0 & 0 & 60 \\ 1 & 1 & 0 & 1 & 0 & 0 & 14 \\ 2 & 1 & 0 & 0 & 1 & 0 & 24 \\ -15 & -15 & 0 & 0 & 0 & 1 & 0 \end{bmatrix}$$
$$\begin{bmatrix} 3/5 & 1 & 1/5 & 0 & 0 & 0 & 12 \\ 2/5 & 0 & -1/5 & 1 & 0 & 0 & 2 \\ 7/5 & 0 & -1/5 & 0 & 1 & 0 & 12 \\ -6 & 0 & 3 & 0 & 0 & 1 & 180 \end{bmatrix}$$

$$\begin{bmatrix} 0 & 1 & 1/2 & -3/2 & 0 & 0 & 9 \\ 1 & 0 & -1/2 & 5/2 & 0 & 0 & 5 \\ 0 & 0 & 1/2 & -7/2 & 1 & 0 & 5 \\ 0 & 0 & 0 & 15 & 0 & 1 & 210 \end{bmatrix}$$
This gives maximum of 210 at (5, 9)

$$\begin{bmatrix} 0 & 1 & 0 & 2 & -1 & 0 & 4 \\ 1 & 0 & 0 & -1 & 1 & 0 & 10 \\ 0 & 0 & 1 & -7 & 2 & 0 & 10 \\ 0 & 0 & 0 & 15 & 0 & 1 & 210 \end{bmatrix}$$
Maximum = 210 is also at (10, 4)

7.
$$\begin{bmatrix} 2 & 1 & 2 & 1 & 0 & 0 & 0 & 20 \\ 1 & 2 & 2 & 0 & 1 & 0 & 0 & 20 \\ 1 & 1 & 4 & 0 & 0 & 1 & 0 & 20 \\ -3 & -3 & -4 & 0 & 0 & 0 & 1 & 0 \end{bmatrix}$$
$$\begin{bmatrix} 3/2 & 1/2 & 0 & 1 & 0 & -1/2 & 0 & 10 \\ 1/2 & 3/2 & 0 & 0 & 1 & -1/2 & 0 & 10 \\ 1/4 & 1/4 & 1 & 0 & 0 & 1/4 & 0 & 5 \\ -2 & -2 & 0 & 0 & 0 & 1 & 1 & 20 \end{bmatrix}$$

$$\begin{bmatrix} 1 & 1/3 & 0 & 2/3 & 0 & -1/3 & 0 & 20/3 \\ 0 & 4/3 & 0 & -1/3 & 1 & -1/3 & 0 & 20/3 \\ 0 & 1/6 & 1 & -1/6 & 0 & 1/3 & 0 & 10/3 \\ 0 & -4/3 & 0 & 4/3 & 0 & 1/3 & 1 & 100/3 \end{bmatrix}$$

$$\begin{bmatrix} 1 & 0 & 0 & 3/4 & -1/4 & -1/4 & 0 & 5 \\ 0 & 1 & 0 & -1/4 & 3/4 & -1/4 & 0 & 5 \\ 0 & 0 & 1 & -1/8 & -1/8 & 3/8 & 0 & 5/2 \\ 0 & 0 & 0 & 1 & 1 & 0 & 1 & 40 \end{bmatrix}$$
Maximum = 40 at (5, 5, 2.5)

$$\begin{bmatrix} 1 & 0 & 2/3 & 2/3 & -1/3 & 0 & 0 & 20/3 \\ 0 & 1 & 2/3 & -1/3 & 2/3 & 0 & 0 & 20/3 \\ 0 & 0 & 8/3 & -1/3 & -1/3 & 1 & 0 & 20/3 \\ 0 & 0 & 0 & 1 & 1 & 0 & 1 & 40 \end{bmatrix}$$

Maximum = 40 also at (20/3, 20/3, 0)

9.
$$\begin{bmatrix} 2 & -5 & 1 & 0 & 0 & 10 \\ 2 & -1 & 0 & 1 & 0 & -2 \\ -8 & -3 & 0 & 0 & 1 & 0 \end{bmatrix} \qquad \begin{bmatrix} -8 & 0 & 1 & -5 & 0 & 20 \\ -2 & 1 & 0 & -1 & 0 & 2 \\ -14 & 0 & 0 & -3 & 1 & 6 \end{bmatrix}$$

Unbounded feasible region

11.
$$\begin{bmatrix} 1 & -3 & 2 & 1 & 0 & 0 & 50 \\ -2 & 4 & 5 & 0 & 1 & 0 & 40 \\ -8 & -6 & -2 & 0 & 0 & 1 & 0 \end{bmatrix} \qquad \begin{bmatrix} 1 & -3 & 2 & 1 & 0 & 0 & 50 \\ 0 & -2 & 9 & 2 & 1 & 0 & 140 \\ 0 & -30 & 14 & 8 & 0 & 1 & 400 \end{bmatrix}$$

Unbounded feasible region

13.
$$\begin{bmatrix} 1 & -1 & 1 & 0 & 0 & -13 \\ 2 & 9 & 0 & 1 & 0 & 72 \\ -12 & -20 & 0 & 0 & 1 & 0 \end{bmatrix} \qquad \begin{bmatrix} -1 & 1 & -1 & 0 & 0 & 13 \\ 11 & 0 & 9 & 1 & 0 & -45 \\ -32 & 0 & -20 & 0 & 1 & 260 \end{bmatrix}$$

Pivoting leads to a row with a negative entry in the last column and all other entries in the row are nonnegative. Thus, there is no feasible solution.

15.
$$\begin{bmatrix} 6 & 4 & 3 & 1 & 0 & 0 & 0 & 60 \\ 3 & 6 & 4 & 0 & 1 & 0 & 0 & 48 \\ 1 & 1 & -2 & 0 & 0 & 1 & 0 & -60 \\ -20 & -30 & -15 & 0 & 0 & 0 & 1 & 0 \end{bmatrix}$$

$$\begin{bmatrix} 15/2 & 11/2 & 0 & 1 & 0 & 3/2 & 0 & -30 \\ 5 & 8 & 0 & 0 & 1 & 2 & 0 & -72 \\ -1/2 & -1/2 & 1 & 0 & 0 & -1/2 & 0 & 30 \\ -55/2 & -75/2 & 0 & 0 & 0 & -15/2 & 1 & 450 \end{bmatrix}$$

Pivoting leads to a row with a negative entry in the last column and all other entries in the row are nonnegative. Thus, there is no feasible solution.

17.
$$\begin{bmatrix} -4 & 1 & 1 & 0 & 0 & 2 \\ 2 & -1 & 0 & 1 & 0 & 1 \\ -1 & -4 & 0 & 0 & 1 & 0 \end{bmatrix} \qquad \begin{bmatrix} -4 & 1 & 1 & 0 & 0 & 2 \\ -2 & 0 & 1 & 1 & 0 & 3 \\ -17 & 0 & 4 & 0 & 1 & 8 \end{bmatrix}$$

Unbounded feasible region since column 1 has all negative entries.

19.
$$\begin{bmatrix} 9 & -4 & -6 & 1 & 0 & 0 & 0 & -36 \\ 4 & 5 & 8 & 0 & 1 & 0 & 0 & 40 \\ 6 & -2 & 1 & 0 & 0 & 1 & 0 & 18 \\ -18 & -15 & -8 & 0 & 0 & 0 & 1 & 0 \end{bmatrix}$$

$$\begin{bmatrix} -3/2 & 2/3 & 1 & -1/6 & 0 & 0 & 0 & 6 \\ 16 & -1/3 & 0 & 4/3 & 1 & 0 & 0 & -8 \\ 15/2 & -8/3 & 0 & 1/6 & 0 & 1 & 0 & 12 \\ -30 & -29/3 & 0 & -4/3 & 0 & 0 & 1 & 48 \end{bmatrix}$$

$$\begin{bmatrix} 61/2 & 0 & 1 & 5/2 & 2 & 0 & 0 & -10 \\ -48 & 1 & 0 & -4 & -3 & 0 & 0 & 24 \\ -241/2 & 0 & 0 & -21/2 & -8 & 1 & 0 & 76 \\ -494 & 0 & 0 & -40 & -29 & 0 & 1 & 280 \end{bmatrix}$$

No feasible solution because row 1 has a negative entry in the last column and nonnegative entries in the rest of the row.

21.
$$\begin{bmatrix} 5 & 3 & 1 & 0 & 0 & 30 \\ -2 & -1 & 0 & 1 & 0 & -20 \\ -15 & -9 & 0 & 0 & 1 & 0 \end{bmatrix} \qquad \begin{bmatrix} 0 & 1/2 & 1 & 5/2 & 0 & -20 \\ 1 & 1/2 & 0 & -1/2 & 0 & 10 \\ 0 & -3/2 & 0 & -15/2 & 1 & 150 \end{bmatrix}$$

No feasible solution because row 1 has a negative entry in the last column and nonnegative entries in the rest of the row.

23.
$$\begin{bmatrix} 10 & 15 & 1 & 0 & 0 & 150 \\ -6 & -3 & 0 & 1 & 0 & -180 \\ -4 & -12 & 0 & 0 & 1 & 0 \end{bmatrix} \qquad \begin{bmatrix} 0 & 10 & 1 & 5/3 & 0 & -150 \\ 1 & 1/2 & 0 & -1/6 & 0 & 30 \\ 0 & -10 & 0 & -2/3 & 1 & 120 \end{bmatrix}$$

No feasible solution because row 1 has a negative entry in the last column and nonnegative entries in the rest of the row.

25.
$$\begin{bmatrix} -2 & -3 & -4 & 1 & 0 & 0 & 0 & -60 \\ -2 & -3 & 6 & 0 & 1 & 0 & 0 & -30 \\ 0 & 6 & -5 & 0 & 0 & 1 & 0 & 30 \\ 2 & 3 & 1 & 0 & 0 & 0 & 1 & 0 \end{bmatrix}$$

$$\begin{bmatrix} 1/2 & 3/4 & 1 & -1/4 & 0 & 0 & 0 & 5 \\ -5 & -15/2 & 0 & 3/2 & 1 & 0 & 0 & -120 \\ 5/2 & 39/4 & 0 & -5/4 & 0 & 1 & 0 & 105 \\ 3/2 & 9/4 & 0 & 1/4 & 0 & 0 & 1 & -15 \end{bmatrix}$$

$$\begin{bmatrix} 0 & 0 & 1 & -1/10 & 1/10 & 0 & 0 & 3 \\ 2/3 & 1 & 0 & -1/5 & -2/15 & 0 & 0 & 16 \\ -4 & 0 & 0 & 7/10 & 13/10 & 1 & 0 & -51 \\ 0 & 0 & 0 & 7/10 & 3/10 & 0 & 1 & -51 \end{bmatrix}$$

$$\begin{bmatrix} 0 & 0 & 1 & -1/10 & 1/10 & 0 & 0 & 3 \\ 0 & 1 & 0 & -1/12 & 1/12 & 1/6 & 0 & 15/2 \\ 1 & 0 & 0 & -7/40 & -13/40 & -1/4 & 0 & 51/4 \\ 0 & 0 & 0 & 7/10 & 3/10 & 0 & 1 & -51 \end{bmatrix}$$

$$\begin{bmatrix} 0 & 0 & 1 & -1/10 & 1/10 & 0 & 0 & 3 \\ 0 & 6 & 0 & -1/2 & 1/2 & 1 & 0 & 45 \\ 1 & 3/2 & 0 & -3/10 & -1/5 & 0 & 0 & 24 \\ 0 & 0 & 0 & 7/10 & 3/10 & 0 & 1 & -51 \end{bmatrix}$$

Multiple solutions, minimum z = 51 at (51/4, 15/2, 3) and (24, 0, 3)

27.
$$\begin{bmatrix} 3 & -10 & 6 & 1 & 0 & 0 & 0 & 60 \\ -15 & 4 & 6 & 0 & 1 & 0 & 0 & 60 \\ 4 & 5 & -20 & 0 & 0 & 1 & 0 & 100 \\ -8 & -5 & -8 & 0 & 0 & 0 & 1 & 0 \end{bmatrix}$$

$$\begin{bmatrix} 1/2 & -5/3 & 1 & 1/6 & 0 & 0 & 0 & 10 \\ -18 & 14 & 0 & -1 & 1 & 0 & 0 & 0 \\ 14 & -85/3 & 0 & 10/3 & 0 & 1 & 0 & 300 \\ -4 & -55/3 & 0 & 4/3 & 0 & 0 & 1 & 80 \end{bmatrix}$$

$$\begin{bmatrix} -23/14 & 0 & 1 & 1/21 & 5/42 & 0 & 0 & 10 \\ -9/7 & 1 & 0 & -1/14 & 1/14 & 0 & 0 & 0 \\ -157/7 & 0 & 0 & 55/42 & 85/42 & 1 & 0 & 300 \\ -193/7 & 0 & 0 & 1/42 & 55/42 & 0 & 1 & 80 \end{bmatrix}$$

Unbounded feasible region

29.   Let x = number of Petite, y = number of Deluxe.
      Maximize z = 5x + 6y   subject to
$$x + 2y \leq 3950$$
$$4x + 3y \leq 9575$$
$$y \geq 2000$$
$$x \geq 0, \; y \geq 0$$

$$\begin{bmatrix} 1 & 2 & 1 & 0 & 0 & 0 & 3950 \\ 4 & 3 & 0 & 1 & 0 & 0 & 9575 \\ 0 & -1 & 0 & 0 & 1 & 0 & -2000 \\ \hline -5 & -6 & 0 & 0 & 0 & 1 & 0 \end{bmatrix}$$

$$\begin{bmatrix} 1 & 0 & 1 & 0 & 2 & 0 & -50 \\ 4 & 0 & 0 & 1 & 3 & 0 & 3575 \\ 0 & 1 & 0 & 0 & -1 & 0 & 2000 \\ \hline -5 & 0 & 0 & 0 & -6 & 1 & 12000 \end{bmatrix}$$

No feasible solution because row 1 has a negative entry in the last column and nonnegative entries in the rest of the row.

31.   Let x = amount invested in stocks and y = amount invested in bonds.
      Maximize expected return
$$z = .10x + .07y \quad \text{subject to}$$
$$x + y \geq 10,000 \text{ (amount invested)}$$
$$x \geq 5000 + y \text{ (5000 more in stocks than bonds)}$$
$$x \geq 0, \; y \geq 0$$
The simplex tableau is

$$\begin{bmatrix} -1 & -1 & 1 & 0 & 0 & -10,000 \\ -1 & 1 & 0 & 1 & 0 & -5,000 \\ \hline -.10 & -.07 & 0 & 0 & 1 & 0 \end{bmatrix}$$

$$\begin{bmatrix} 1 & 1 & -1 & 0 & 0 & 10,000 \\ 0 & 2 & -1 & 1 & 0 & 5,000 \\ \hline 0 & .03 & -.1 & 0 & 1 & 1,000 \end{bmatrix}$$

Column 3 contains all negative entries which indicates an unbounded feasible region so there is no maximum expected return.

33.   Let x_1 = number units of food A
           x_2 = number units of food B
           x_3 = number units of food C
Minimize cost: $z = 1.40x_1 + x_2 + 1.50x_3$ subject to
$$15x_1 + 10x_2 + 23x_3 \geq 80 \text{ (protein)}$$
$$20x_1 + 30x_2 + 11x_3 \geq 95 \text{ (carbohydrates)}$$
$$500x_1 + 400x_2 + 400x_3 \geq 3500 \text{ (calories)}$$
$$10x_1 + 5x_2 + 6x_3 \leq 40 \text{ (fat)}$$
$$x_1 \geq 0, \; x_2 \geq 0, \; x_3 \geq 0$$

$$\begin{bmatrix} -15 & -10 & -23 & 1 & 0 & 0 & 0 & 0 & -80 \\ -20 & -30 & -11 & 0 & 1 & 0 & 0 & 0 & -95 \\ -500 & -400 & -400 & 0 & 0 & 1 & 0 & 0 & -3500 \\ 10 & 5 & 6 & 0 & 0 & 0 & 1 & 0 & 40 \\ \hline 1.4 & 1.0 & 1.5 & 0 & 0 & 0 & 0 & 1 & 0 \end{bmatrix}$$

$$\begin{bmatrix} 0 & 2.0 & -11 & 1 & 0 & -.03 & 0 & 0 & 25 \\ 0 & -14.0 & 5 & 0 & 1 & -.04 & 0 & 0 & 45 \\ 1 & .8 & .8 & 0 & 0 & 0 & 0 & 0 & 7 \\ 0 & -3.0 & -2 & 0 & 0 & .02 & 1 & 0 & -30 \\ 0 & -.12 & .38 & 0 & 0 & 0 & 0 & 1 & -9.8 \end{bmatrix}$$

$$\begin{bmatrix} 0 & 0 & -12.33 & 1 & 0 & -.02 & .67 & 0 & 5 \\ 0 & 0 & 14.33 & 0 & 1 & -.13 & -4.67 & 0 & 185 \\ 1 & 0 & .27 & 0 & 0 & 0 & .27 & 0 & -1 \\ 0 & 1 & .67 & 0 & 0 & -.01 & -.33 & 0 & 10 \\ 0 & 0 & .46 & 0 & 0 & .002 & -.04 & 1 & -8.6 \end{bmatrix}$$

Notice row 3. The last entry is negative and all other entries are nonnegative. This indicates no feasible solution.

Section 4.6

1. (a) For $(0, 0)$ $s_1 = 17$. For $(2, 2)$ $s_1 = 3$. For $(5, 10)$ $s_1 = -33$. For $(2, 1)$ $s_1 = 6$. For $(2, 3)$ $s_1 = 0$.
   (b) $(0, 0)$, $(2, 2)$, $(2, 1)$, and $(2, 3)$ are in the feasible region.
   (c) $(2, 3)$ is on the line.

3. Let $x_2 = 0$, $s_2 = 0$ in the second constraint to obtain $5x_1 = 30$, $x_1 = 6$. Thus, the corner point is $(6, 0)$.

5. For $(0, 0, 0)$, $s_1 = 40$. For $(1, 2, 3)$, $s_1 = 26$.
   For $(0, 10, 0)$, $s_1 = 0$. For $(4, 2, 7)$, $s_1 = 13$.

7.

Point	s_1	s_2	s_3	Is point on boundary?	Is point in feasible region?
$(5, 10)$	15	32	18	No	Yes
$(8, 10)$	12	14	-3	No	No
$(5, 13)$	0	29	0	Yes	Yes
$(11, 13)$	-6	-7	-42	No	No
$(10, 12)$	0	0	-29	No	No
$(15, 11)$	0	-29	-58	No	No

9. (i) (a) $x_1 = 0$, $x_2 = 0$, $s_1 = 900$, $s_2 = 2800$, $z = 0$
   (b) Intersection of $x_1 = 0$ and $x_2 = 0$
   (ii) (a) $x_1 = 180$, $x_2 = 0$, $s_1 = 0$, $s_2 = 1360$, $z = 540$
   (b) Intersection of $5x_1 + 2x_2 = 900$ and $x_2 = 0$
   (iii) (a) $x_1 = 100$, $x_2 = 200$, $s_1 = 0$, $s_2 = 0$, $z = 700$
   (b) Intersection of $5x_1 + 2x_2 = 900$ and $8x_1 + 10x_2 = 2800$

11. (a) For $(3, 5)$ to be on the boundary line $s_1 = 0$ and $6(3) + 4(5) = 24$. Since the latter is false, $(3, 5)$ is not on the boundary line.
    (b) For $(3, 5)$ to be in the feasible region, $6(3) + 4(5) + s_1 = 24$ for some nonnegative value of s_1. However, the equation is true only when $s_1 = -14$ so $(3, 5)$ cannot be in the feasible region.

13.   (a) When $x_1 = 0$, $\quad 7x_2 + s_1 = 36$ and $5x_2 + s_2 = 32$.

             If $s_1 \geq 0$, $\quad 36 \geq 7x_2$ and $x_2 \leq 36/7$.

             If $s_2 \geq 0$, $\quad 32 \geq 5x_2$ and $x_2 \leq 32/5$.

             Both $x_2 \leq 36/7$ and $x_2 \leq 32/5$ must hold, so x_2 must be the smaller, 36/7.

    (b) When $x_2 = 0$, $\quad 6x_1 + s_1 = 36$ and $2x_1 + s_2 = 32$.

             If $s_1 \geq 0$, $\quad 36 \geq 6x_1$ and $x_1 \leq 6$.

             If $s_2 \geq 0$, $\quad 32 \geq 2x_2$ and $x_2 \leq 16$.

             Both $x_1 \leq 6$ and $x_1 \leq 16$ must hold, so x_1 is the smaller, 6.

15.   Initial tableau

$$\begin{bmatrix} -7 & 10 & 1 & 0 & 0 & 0 & 0 & 50 \\ 3 & 5 & 0 & 1 & 0 & 0 & 0 & 90 \\ 4 & 5 & 0 & 0 & 1 & 0 & 0 & 105 \\ 1 & 0 & 0 & 0 & 0 & 1 & 0 & 20 \\ \hline -17 & -20 & 0 & 0 & 0 & 0 & 1 & 0 \end{bmatrix}$$

Corner $(0, 0)$, $z = 0$,
slack $(50, 90, 105, 20)$

Tableau 1

$$\begin{bmatrix} -.7 & 1 & .1 & 0 & 0 & 0 & 0 & 5 \\ 6.5 & 0 & -.5 & 1 & 0 & 0 & 0 & 65 \\ 7.5 & 0 & -.5 & 0 & 1 & 0 & 0 & 80 \\ 1.0 & 0 & 0 & 0 & 0 & 1 & 0 & 20 \\ \hline -31 & 0 & 2 & 0 & 0 & 0 & 1 & 100 \end{bmatrix}$$

Corner $(0, 5)$, $z = 100$,
slack $(0, 65, 80, 20)$

Tableau 2

$$\begin{bmatrix} 0 & 1 & .046 & .108 & 0 & 0 & 0 & 12 \\ 1 & 0 & -.077 & .154 & 0 & 0 & 0 & 10 \\ 0 & 0 & .077 & -1.154 & 1 & 0 & 0 & 5 \\ 0 & 0 & .077 & -.154 & 0 & 1 & 0 & 10 \\ \hline 0 & 0 & -.385 & 4.769 & 0 & 0 & 1 & 410 \end{bmatrix}$$

Corner $(10, 12)$, $z = 410$, slack $(0, 0, 5, 10)$

Tableau 3

$$\begin{bmatrix} 0 & 1 & 0 & .80 & -.6 & 0 & 0 & 9 \\ 1 & 0 & 0 & -1.00 & 1.0 & 0 & 0 & 15 \\ 0 & 0 & 1 & -15.00 & 13.0 & 0 & 0 & 65 \\ 0 & 0 & 0 & 1.00 & -1.0 & 1 & 0 & 5 \\ \hline 0 & 0 & 0 & -1.00 & 5.0 & 0 & 1 & 435 \end{bmatrix}$$

Corner $(15, 9)$, $z = 435$,
slack $(65, 0, 0, 5)$

Tableau 4

$$\begin{bmatrix} 0 & 1 & 0 & 0 & .2 & -.8 & 0 & 5 \\ 1 & 0 & 0 & 0 & 0 & 1.0 & 0 & 20 \\ 0 & 0 & 1 & 0 & -2.0 & 15.0 & 0 & 140 \\ 0 & 0 & 0 & 1 & -1.0 & 1.0 & 0 & 5 \\ \hline 0 & 0 & 0 & 0 & 4.0 & 1.0 & 1 & 440 \end{bmatrix}$$

Optimal solution reached
Corner $(20, 5)$, $z = 440$, slack $(140, 5, 0, 0)$

Corner (x,y)	Value of z	Increase in z	Slack (s_1, s_2, s_3, s_4)
$(0, 0)$	0		$(50, 90, 105, 20)$
$(0, 5)$	100	100	$(0, 65, 80, 20)$
$(10, 12)$	410	310	$(0, 0, 5, 10)$
$(15, 9)$	435	25	$(65, 0, 0, 5)$
$(20, 5)$	440	5	$(140, 5, 0, 0)$

17. Initial tableau

$$\begin{bmatrix} 1 & 6 & 3 & 1 & 0 & 0 & 0 & | & 36 \\ 3 & 6 & 6 & 0 & 1 & 0 & 0 & | & 45 \\ 5 & 6 & 1 & 0 & 0 & 1 & 0 & | & 46 \\ -10 & -24 & -13 & 0 & 0 & 0 & 1 & | & 0 \end{bmatrix}$$

Corner (0, 0), z = 0 slack (36, 45, 46)

Tableau 1

$$\begin{bmatrix} .167 & 1 & .5 & .167 & 0 & 0 & | & 6 \\ 2.000 & 0 & 3.0 & -1.000 & 1 & 0 & | & 9 \\ 4.000 & 0 & -2.0 & -1.000 & 0 & 1 & | & 10 \\ -6.000 & 0 & -1.0 & 4.000 & 0 & 0 & | & 144 \end{bmatrix}$$

Corner (0, 6, 0), z = 144, slack (0, 9, 10)

Tableau 2

$$\begin{bmatrix} 0 & 1 & .583 & .208 & 0 & -.042 & 0 & | & 5.583 \\ 0 & 0 & 4.000 & -.500 & 1 & -.500 & 0 & | & 4.000 \\ 1 & 0 & -.500 & -.250 & 0 & .250 & 0 & | & 2.500 \\ 0 & 0 & -4.000 & 2.500 & 0 & 1.500 & 1 & | & 159 \end{bmatrix}$$

Corner (2.5, 5.583, 0), z = 159, slack (0, 4, 0)

Tableau 3

$$\begin{bmatrix} 0 & 1 & 0 & .281 & -.146 & .031 & 0 & | & 5 \\ 0 & 0 & 1 & -.125 & .250 & -.125 & 0 & | & 1 \\ 1 & 0 & 0 & -.313 & .125 & .188 & 0 & | & 3 \\ 0 & 0 & 0 & 2.000 & 1.000 & 1.000 & 1 & | & 163 \end{bmatrix}$$

Corner (3, 5, 1), z = 163, slack (0, 0, 0) Optimal solution

Corner (x_1, x_2, x_3)	Value of z	Increase in z	Slack (s_1, s_2, s_3)
(0, 0, 0)	0		(36, 45, 46)
(0, 6, 0)	144	144	(0, 9, 10)
(2.5, 5.583, 0)	159	15	(0, 4, 0)
(3, 5, 1)	163	4	(0, 0, 0)

19. Initial Tableau

$$\begin{bmatrix} 2 & 1 & 2 & 1 & 0 & 0 & 0 & | & 330 \\ 1 & 2 & 2 & 0 & 1 & 0 & 0 & | & 330 \\ -2 & -2 & 1 & 0 & 0 & 1 & 0 & | & 132 \\ -1 & -2 & -3 & 0 & 0 & 0 & 1 & | & 0 \end{bmatrix}$$

Corner (0, 0, 0), z = 0, slack (330, 330, 132)

Tableau 1

$$\begin{bmatrix} 6 & 5 & 0 & 1 & 0 & -2 & 0 & | & 66 \\ 5 & 6 & 0 & 0 & 1 & -2 & 0 & | & 66 \\ -2 & -2 & 1 & 0 & 0 & 1 & 0 & | & 132 \\ -7 & -8 & 0 & 0 & 0 & 3 & 1 & | & 396 \end{bmatrix}$$

Corner (0, 0, 132), z = 396, slack (66, 66, 0)

Tableau 2

$$\begin{bmatrix} 1.833 & 0 & 0 & 1 & -.833 & -.333 & 0 & | & 11 \\ .833 & 1 & 0 & 0 & .167 & -.333 & 0 & | & 11 \\ -.333 & 0 & 1 & 0 & .333 & .333 & 0 & | & 154 \\ -.333 & 0 & 0 & 0 & 1.333 & .333 & 1 & | & 484 \end{bmatrix}$$

105

Corner (0, 11, 154), z = 484, slack (11, 0, 0)

Tableau 3

$$\begin{bmatrix} 1 & 0 & 0 & .545 & -.454 & -.182 & 0 & 6 \\ 0 & 1 & 0 & -.454 & .545 & -.182 & 0 & 6 \\ 0 & 0 & 1 & .182 & .182 & .273 & 0 & 156 \\ \hline 0 & 0 & 0 & .182 & .182 & .273 & 1 & 486 \end{bmatrix}$$

Corner (6, 6, 156), z = 486, slack (0, 0, 0)

Corner (x_1, x_2, x_3)	Value of z	Increase in z	Slack (s_1, s_2, s_3)
(0, 0, 0)	0		(330, 330, 132)
(0, 0, 132)	396	396	(66, 66, 0)
(0, 11, 154)	484	88	(11, 0, 0)
(6, 6, 156)	486	2	(0, 0, 0)

21. There is no change in z because the the process remains at that corner point but changes from one boundary line to another at that point.

Review Exercises, Chapter 4

1. $6x_1 + 4x_2 + 3x_3 + s_1 = 220$
   $x_1 + 5x_2 + x_3 + s_2 = 162$
   $7x_1 + 2x_2 + 5x_3 + s_3 = 139$

3. $6x_1 + 5x_2 + 3x_3 + 3x_4 + s_1 = 89$
   $7x_1 + 4x_2 + 6x_3 + 2x_4 + s_2 = 72$

5. $10x_1 + 12x_2 + 8x_3 + s_1 = 24$
   $7x_1 + 13x_2 + 5x_3 + s_2 = 35$
   $-20x_1 - 36x_2 - 19x_3 + z = 0$

7. $3x_1 + 7x_2 + s_1 = 14$
   $9x_1 + 5x_2 + s_2 = 18$
   $x_1 - x_2 + s_3 = 21$
   $-9x_1 - 2x_2 + z = 0$

9. $x_1 + x_2 + x_3 + s_1 = 15$
   $2x_1 + 4x_2 + x_3 + s_2 = 44$
   $-6x_1 - 8x_2 - 4x_3 + z = 0$

11. (a)
$$\begin{bmatrix} 5 & 3 & 2 & 1 & 0 & 0 & 0 & 600 \\ 4 & 6 & 1 & 0 & 1 & 0 & 0 & 900 \\ 1 & 2 & 3 & 0 & 0 & 1 & 0 & 800 \\ -5 & -8 & -4 & 0 & 0 & 0 & 1 & 0 \end{bmatrix}$$
$600/3 = 200$
$900/6 = 150$ Pivot Row
$800/2 = 400$

Column 2 is the pivot column.
The 6 in row 2, column 2 is the pivot element.

(b)
$$\begin{bmatrix} 1 & 4 & 0 & 3 & 0 & -2 & 0 & 60 \\ 0 & 6 & 1 & 5 & 0 & 4 & 0 & 60 \\ 0 & -3 & 0 & 1 & 1 & 2 & 0 & 60 \\ 0 & -1 & 0 & -2 & 0 & 3 & 1 & 0 \end{bmatrix}$$
$60/3 = 20$
$60/5 = 12$ Pivot Row
$60/1 = 60$

Column 4 is pivot column.

Review, Chapter 4

The 5 in row 2, column 4 is the pivot element.

13. (a) $x_1 = 0_1$, $x_2 = 80$, $s_1 = 0$, $s_2 = 42$, $z = 98$

(b) $x_1 = 73$, $x_2 = 42$, $x_3 = 15$, $s_1 = 0$, $s_2 = 0$, $s_3 = 0$,
$z = 138$

15. (a)
$$\begin{bmatrix} 11 & 5 & 3 & 1 & 0 & 0 & 0 & 142 \\ -3 & -4 & -7 & 0 & 1 & 0 & 0 & -95 \\ 2 & 15 & 1 & 0 & 0 & 1 & 0 & 124 \\ -3 & -5 & -4 & 0 & 0 & 0 & 1 & 0 \end{bmatrix}$$

(b)
$$\begin{bmatrix} 7 & 4 & 1 & 0 & 0 & 28 \\ -1 & -3 & 0 & 1 & 0 & -6 \\ 14 & 22 & 0 & 0 & 1 & 0 \end{bmatrix}$$

17. (a)
$$\begin{bmatrix} -15 & -8 & 1 & 0 & 0 & 0 & 0 & -120 \\ 10 & 12 & 0 & 1 & 0 & 0 & 0 & 120 \\ 15 & 5 & 0 & 0 & 1 & 0 & 0 & 75 \\ -15 & -5 & 0 & 0 & 0 & 1 & 0 & -75 \\ -5 & -12 & 0 & 0 & 0 & 0 & 1 & 0 \end{bmatrix}$$

(b)
$$\begin{bmatrix} 14 & 9 & 1 & 0 & 0 & 0 & 0 & 126 \\ -10 & -11 & 0 & 1 & 0 & 0 & 0 & -110 \\ -5 & 1 & 0 & 0 & 1 & 0 & 0 & 9 \\ 5 & -1 & 0 & 0 & 0 & 1 & 0 & -9 \\ 3 & 2 & 0 & 0 & 0 & 0 & 1 & 0 \end{bmatrix}$$

19.
$$\begin{bmatrix} 2 & 4 & 2 & 1 & 0 & 0 & 0 & 34 \\ 3 & 6 & 4 & 0 & 1 & 0 & 0 & 57 \\ 2 & 5 & 1 & 0 & 0 & 1 & 0 & 30 \\ -3 & -5 & -2 & 0 & 0 & 0 & 1 & 0 \end{bmatrix} \begin{bmatrix} 2/5 & 0 & 6/5 & 1 & 0 & -4/5 & 0 & 10 \\ 3/5 & 0 & 14/5 & 0 & 1 & -6/5 & 0 & 21 \\ 2/5 & 1 & 1/5 & 0 & 0 & 1/5 & 0 & 6 \\ -1 & 0 & -1 & 0 & 0 & 1 & 1 & 30 \end{bmatrix}$$

$$\begin{bmatrix} 0 & -1 & 1 & 1 & 0 & -1 & 0 & 4 \\ 0 & -3/2 & 5/2 & 0 & 1 & -3/2 & 0 & 12 \\ 1 & 5/2 & 1/2 & 0 & 0 & 1/2 & 0 & 15 \\ 0 & 5/2 & -1/2 & 0 & 0 & 3/2 & 1 & 45 \end{bmatrix}$$

$$\begin{bmatrix} 0 & -1 & 1 & 1 & 0 & -1 & 0 & 4 \\ 0 & 1 & 0 & -5/2 & 1 & 1 & 0 & 2 \\ 1 & 3 & 0 & -1/2 & 0 & 1 & 0 & 13 \\ 0 & 2 & 0 & 1/2 & 0 & 1 & 1 & 47 \end{bmatrix}$$

Maximum $z = 47$ at $(13, 0, 4)$

21.
$$\begin{bmatrix} -4 & 1 & 1 & 0 & 0 & 3 \\ 1 & -2 & 0 & 1 & 0 & 12 \\ -10 & -15 & 0 & 0 & 1 & 0 \end{bmatrix} \begin{bmatrix} -4 & 1 & 1 & 0 & 0 & 3 \\ -7 & 0 & 2 & 1 & 0 & 18 \\ -70 & 0 & 15 & 0 & 1 & 45 \end{bmatrix}$$

Unbounded feasible region, no maximum.

23.
$$\begin{bmatrix} 4 & 1 & 1 & 1 & 0 & 0 & 372 \\ 1 & 8 & 6 & 0 & 1 & 0 & 1116 \\ -1 & -3 & -1 & 0 & 0 & 1 & 0 \end{bmatrix}$$

$$\begin{bmatrix} 31/8 & 0 & 1/4 & 1 & -1/8 & 0 & 465/2 \\ 1/8 & 1 & 3/4 & 0 & 1/8 & 0 & 279/2 \\ -5/8 & 0 & 5/4 & 0 & 3/8 & 1 & 837/2 \end{bmatrix}$$

$$\begin{bmatrix} 1 & 0 & 2/31 & 8/31 & -1/31 & 0 & 60 \\ 0 & 1 & 23/31 & -1/31 & 4/31 & 0 & 132 \\ 0 & 0 & 40/31 & 5/31 & 11/31 & 1 & 456 \end{bmatrix}$$ Maximum $z = 456$ at $(60,132,0)$

25. (a) x_3 (b) x_2

27.
$$\begin{bmatrix} 3 & 4 & 5 \\ 1 & 0 & 7 \\ -2 & 6 & 8 \end{bmatrix} \qquad \begin{bmatrix} 4 & -5 \\ 3 & 0 \\ 2 & 12 \\ 1 & 9 \end{bmatrix}$$

29.
$$\left[\begin{array}{cccccc|c} -2 & -1 & 0 & 1 & 0 & 0 & -6 \\ 0 & -1 & -2 & 0 & 1 & 0 & -8 \\ 6 & 8 & 16 & 0 & 0 & 1 & 0 \end{array}\right] \qquad \left[\begin{array}{cccccc|c} 1 & 1/2 & 0 & -1/2 & 0 & 0 & 3 \\ 0 & -1 & -2 & 0 & 1 & 0 & -8 \\ 0 & 5 & 16 & 3 & 0 & 1 & -18 \end{array}\right]$$

$$\left[\begin{array}{cccccc|c} 1 & 1/2 & 0 & -1/2 & 0 & 0 & 3 \\ 0 & 1/2 & 1 & 0 & -1/2 & 0 & 4 \\ 0 & -3 & 0 & 3 & 8 & 1 & -82 \end{array}\right]$$

$$\left[\begin{array}{cccccc|c} 2 & 1 & 0 & -1 & 0 & 0 & 6 \\ -1 & 0 & 1 & 1/2 & -1/2 & 0 & 1 \\ 6 & 0 & 0 & 0 & 8 & 1 & -64 \end{array}\right] \qquad \left[\begin{array}{cccccc|c} 0 & 1 & 2 & 0 & -1 & 0 & 8 \\ -2 & 0 & 2 & 1 & -1 & 0 & 2 \\ 6 & 0 & 0 & 0 & 8 & 1 & -64 \end{array}\right]$$

Multiple solutions, minimum z = 64 at (0, 6, 1) and (0, 8, 0)

31.
$$\left[\begin{array}{ccccccc|c} 2 & 5 & 4 & 1 & 0 & 0 & 0 & 40 \\ 40 & 45 & 30 & 0 & 1 & 0 & 0 & 430 \\ -6 & -3 & -4 & 0 & 0 & 1 & 0 & -48 \\ -6 & -11 & -8 & 0 & 0 & 0 & 1 & 0 \end{array}\right] \quad \left[\begin{array}{ccccccc|c} 0 & 4 & 8/3 & 1 & 0 & 1/3 & 0 & 24 \\ 0 & 25 & 10/3 & 0 & 1 & 20/3 & 0 & 110 \\ 1 & 1/2 & 2/3 & 0 & 0 & -1/6 & 0 & 8 \\ 0 & -8 & -4 & 0 & 0 & -1 & 1 & 48 \end{array}\right]$$

$$\left[\begin{array}{ccccccc|c} 0 & 0 & 32/15 & 1 & -4/25 & -11/15 & 0 & 32/5 \\ 0 & 1 & 2/15 & 0 & 1/25 & 4/15 & 0 & 22/5 \\ 1 & 0 & 3/5 & 0 & -1/50 & -3/10 & 0 & 29/5 \\ 0 & 0 & -44/15 & 0 & 8/25 & 17/15 & 1 & 416/5 \end{array}\right]$$

$$\left[\begin{array}{ccccccc|c} 0 & 0 & 1 & 15/32 & -3/40 & -11/32 & 0 & 3 \\ 0 & 1 & 0 & -1/16 & 1/20 & 5/16 & 0 & 4 \\ 1 & 0 & 0 & -9/32 & 1/40 & -3/32 & 0 & 4 \\ 0 & 0 & 0 & 11/8 & 1/10 & 1/8 & 1 & 92 \end{array}\right]$$

Maximum z = 92 at (4, 4, 3)

33.
$$\left[\begin{array}{ccccccc|c} -1 & -1 & -1 & 1 & 0 & 0 & 0 & -6 \\ 2 & 1 & 3 & 0 & 1 & 0 & 0 & 10 \\ 0 & 2 & -1 & 0 & 0 & 1 & 0 & 5 \\ -2 & -5 & -3 & 0 & 0 & 0 & 1 & 0 \end{array}\right] \quad \left[\begin{array}{ccccccc|c} 1 & 1 & 1 & -1 & 0 & 0 & 0 & 6 \\ 0 & -1 & 1 & 2 & 1 & 0 & 0 & -2 \\ 0 & 2 & -1 & 0 & 0 & 1 & 0 & 5 \\ 0 & -3 & -1 & -2 & 0 & 0 & 1 & 12 \end{array}\right]$$

$$\left[\begin{array}{ccccccc|c} 1 & 0 & 2 & 1 & 1 & 0 & 0 & 4 \\ 0 & 1 & -1 & -2 & -1 & 0 & 0 & 2 \\ 0 & 0 & 1 & 4 & 2 & 1 & 0 & 1 \\ 0 & 0 & -4 & -8 & -3 & 0 & 1 & 18 \end{array}\right]$$

$$\left[\begin{array}{ccccccc|c} 1 & 0 & 7/4 & 0 & 1/2 & -1/4 & 0 & 15/4 \\ 0 & 1 & -1/2 & 0 & 0 & 1/2 & 0 & 5/2 \\ 0 & 0 & 1/4 & 1 & 1/2 & 1/4 & 0 & 1/4 \\ 0 & 0 & -2 & 0 & 1 & 2 & 1 & 20 \end{array}\right]$$

$$\left[\begin{array}{ccccccc|c} 1 & 0 & 0 & -7 & -3 & -2 & 0 & 2 \\ 0 & 1 & 0 & 2 & 1 & 1 & 0 & 3 \\ 0 & 0 & 1 & 4 & 2 & 1 & 0 & 1 \\ 0 & 0 & 0 & 8 & 5 & 4 & 1 & 22 \end{array}\right] \qquad \text{Maximum z = 22 at (2, 3, 1)}$$

35.
$$
\begin{bmatrix}
1 & -3 & 1 & 0 & 0 & | & 24 \\
-5 & 4 & 0 & 1 & 0 & | & 20 \\
-20 & -32 & 0 & 0 & 1 & | & 0
\end{bmatrix}
\quad
\begin{bmatrix}
-11/4 & 0 & 1 & 3/4 & 0 & | & 39 \\
-5/4 & 1 & 0 & 1/4 & 0 & | & 5 \\
-60 & 0 & 0 & 8 & 1 & | & 160
\end{bmatrix}
$$

Unbounded feasible region, no maximum.

37.
$$
\begin{bmatrix}
-3 & -2 & 1 & 0 & 0 & 0 & | & -24 \\
-5 & -4 & 0 & 1 & 0 & 0 & | & -46 \\
-4 & -9 & 0 & 0 & 1 & 0 & | & -60 \\
18 & 36 & 0 & 0 & 0 & 1 & | & 0
\end{bmatrix}
\quad
\begin{bmatrix}
1 & 2/3 & -1/3 & 0 & 0 & 0 & | & 8 \\
0 & -2/3 & -5/3 & 1 & 0 & 0 & | & -6 \\
0 & -19/3 & -4/3 & 0 & 1 & 0 & | & -28 \\
0 & 24 & 6 & 0 & 0 & 1 & | & -144
\end{bmatrix}
$$

$$
\begin{bmatrix}
1 & 0 & -2 & 1 & 0 & 0 & | & 2 \\
0 & 1 & 5/2 & -3/2 & 0 & 0 & | & 9 \\
0 & 0 & 29/2 & -19/2 & 1 & 1 & | & 29 \\
0 & 0 & -54 & 36 & 0 & 0 & | & -360
\end{bmatrix}
$$

$$
\begin{bmatrix}
1 & 0 & 0 & -9/29 & 4/29 & 0 & | & 6 \\
0 & 1 & 0 & 4/29 & -5/29 & 0 & | & 4 \\
0 & 0 & 1 & -19/29 & 2/29 & 0 & | & 2 \\
0 & 0 & 0 & 18/29 & 108/29 & 1 & | & -252
\end{bmatrix}
$$
Minimum $z = 252$ at $(6, 4)$

39.
$$
\begin{bmatrix}
4 & 1 & 1 & 0 & 0 & 0 & 0 & | & 180 \\
-1 & -3 & 0 & 1 & 0 & 0 & 0 & | & -120 \\
-1 & 3 & 0 & 0 & 1 & 0 & 0 & | & 150 \\
1 & -3 & 0 & 0 & 0 & 1 & 0 & | & -150 \\
-5 & -15 & 0 & 0 & 0 & 0 & 1 & | & 0
\end{bmatrix}
$$

$$
\begin{bmatrix}
11/3 & 0 & 1 & 1/3 & 0 & 0 & 0 & | & 140 \\
1/3 & 1 & 0 & -1/3 & 0 & 0 & 0 & | & 40 \\
-2 & 0 & 0 & 1 & 1 & 0 & 0 & | & 30 \\
2 & 0 & 0 & -1 & 0 & 1 & 0 & | & -30 \\
0 & 0 & 0 & -5 & 0 & 0 & 1 & | & 600
\end{bmatrix}
$$

$$
\begin{bmatrix}
13/3 & 0 & 1 & 0 & 0 & 1/3 & 0 & | & 130 \\
-1/3 & 1 & 0 & 0 & 0 & -1/3 & 0 & | & 50 \\
0 & 0 & 0 & 0 & 1 & 1 & 0 & | & 0 \\
-2 & 0 & 0 & 1 & 0 & -1 & 0 & | & 30 \\
-10 & 0 & 0 & 0 & 0 & -5 & 1 & | & 750
\end{bmatrix}
$$

$$
\begin{bmatrix}
1 & 0 & 3/13 & 0 & 0 & 1/13 & 0 & | & 30 \\
0 & 1 & 1/13 & 0 & 0 & -4/13 & 0 & | & 60 \\
0 & 0 & 0 & 0 & 1 & 1 & 0 & | & 0 \\
0 & 0 & 6/13 & 1 & 0 & -11/13 & 0 & | & 90 \\
0 & 0 & 30/13 & 0 & 0 & -55/13 & 1 & | & 1050
\end{bmatrix}
$$

$$
\begin{bmatrix}
1 & 0 & 3/13 & 0 & -1/13 & 0 & 0 & | & 30 \\
0 & 1 & 1/13 & 0 & 4/13 & 0 & 0 & | & 60 \\
0 & 0 & 0 & 0 & 1 & 1 & 0 & | & 0 \\
0 & 0 & 6/13 & 1 & 11/13 & 0 & 0 & | & 90 \\
0 & 0 & 30/13 & 0 & 55/13 & 0 & 1 & | & 1050
\end{bmatrix}
$$

Maximum $z = 1050$ at $(30, 60)$

41. $\begin{bmatrix} 1 & 3 & 1 & 0 & 0 & | & 9 \\ 1 & -1 & 0 & 1 & 0 & | & -2 \\ -2 & -1 & 0 & 0 & 1 & | & 0 \end{bmatrix}$ $\begin{bmatrix} 4 & 0 & 1 & 3 & 0 & | & 3 \\ -1 & 1 & 0 & -1 & 0 & | & 2 \\ -3 & 0 & 0 & -1 & 1 & | & 2 \end{bmatrix}$

$\begin{bmatrix} 1 & 0 & 1/4 & 3/4 & 0 & | & 3/4 \\ 0 & 1 & 1/4 & -1/4 & 0 & | & 11/4 \\ 0 & 0 & 3/4 & 5/4 & 1 & | & 17/4 \end{bmatrix}$ Maximum $z = 17/4$ at $(3/4, 11/4)$

43.   Let x_1 = number of hunting jackets, x_2 = number of
      all-weather jackets, x_3 = number of ski jackets.
      Maximize $z = 7.5x_1 + 9x_2 + 11x_3$ subject to

$$3x_1 + 2.5x_2 + 3.5x_3 \leq 3200 \quad \text{(Labor)}$$
$$26x_1 + 20x_2 + 22x_3 \leq 18000 \quad \text{(Operating costs)}$$
$$x_1 \geq 0, \ x_2 \geq 0, \ x_3 \geq 0$$

$\begin{bmatrix} 3 & 2.5 & 3.5 & 1 & 0 & 0 & | & 3200 \\ 26 & 20 & 22 & 0 & 1 & 0 & | & 18000 \\ -7.5 & -9 & -11 & 0 & 0 & 1 & | & 0 \end{bmatrix}$

Chapter 5
Sets and Counting

Section 5.1

1. (a) True   (b) False   (c) False   (d) False
    (e) True   (f) True   (g) False   (h) True
    (i) False

3. B = {M, I, S, P}   5.   C = {16, 18, 20, 22, ...}

7. Not equal   9.   Equal   11. Not equal

13. A ⊂ B   15.   A ⊄ B   17.   A ⊂ B

19. A ⊂ B

21. (a) ∅, {-1}, {2}, {4}, {-1, 2}, {-1, 4}, {2, 4}, {-1, 2, 4}
     (b) ∅, {4}
     (c) ∅, {-3}, {5}, {6}, {8}, {-3, 5}, {-3, 6}, {-3, 8}, {5, 6},
          {5, 8}, {6, 8}, {-3, 5, 6}, {-3, 5, 8}, {-3, 6, 8},
          {6, 5, 8}, {-3, 5, 6, 8}

23. ∅   25.   Not empty   27.   ∅

29. ∅   31.   ∅

33. {1, 2, 4, 6, 7}   35.   {a, b, c, d, x, y, z}

37. {9, 12}

39. A' = {17, 18, 19}   41.   A' = {11, 12, 13}

43. (a) A' = {-1, 0, 12}   (b) B' = {1, 12, 13}
     (c) (A ∪ B)' = {12}   (d) (A ∩ B)' = {-1, 0, 1, 12, 13}

45. A ∩ B = {1, 2, 3}   47.   A ∪ B = {1, 2, 3, 6, 9}

49. A ∩ B ∩ C = {2, 3}   51.   A ∩ ∅ = ∅

53. (A ∩ C) ∪ B = {1, 2, 3, 6, 9}

55. {13, 22, 33, ...}   57.   {1, 2, 3, 4}

59. (a) ⊂   (b) Neither (c) =   (d) Neither   (e) ⊂

61. A ∩ B = {x|x is an integer that is a multiple of 35}

63. Disjoint

65. A ∩ B is the set of students at Miami Bay University who are
     taking both finite math and American history.

67. (a) A ∩ B is the set of students at Winfield College who had a
          4.0 GPA for both the Fall, 1997 and Spring, 1998 semester.
     (b) A is not necessarily a subset of B because some students may
          have a 4.0 GPA in the fall semester and not in the spring
          semester.

Section 5.2

1. (a) n(A) = 10   (b)   n(B) = 5   (c) n(A ∩ B) = 4
    (d) n(A ∪ B) = 10 + 5 - 4 = 11

3. n(A ∪ B) = 120 + 100 - 40 = 180

5. $30 = 15 + 22 - n(A \cap B)$ $n(A \cap B) = 7$

7. $n(B) = 28 - 14 + 5 = 19$ 9. $60 + 75 - 100 = 35$

11.   (a) 42   (b) 60   (c) 84   (d) 57
       (e) 39   (f) 81   (g) 15   (h) 15
       (i) 81

13.   (a) 35   (b) 9   (c) 6   (d) 38
       (e) 38   (f) 12

15.   From the given information we can fill in the
       regions shown in the Venn diagram.

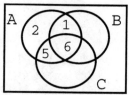

       We can also compute $n(A \cup B) = 14 + 20 - 7 = 27$; $n(A \cup C) = 14 + 23 - 11 = 26$. Since $n(A \cup B \cup C) = 33$ and $n(A \cup B) = 27$, the number in C alone is $33 - 27 = 6$. Likewise, the number in B alone is $33 - 26 = 7$. This gives the Venn diagram

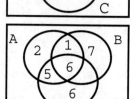

       The one remaining region must be 6 in order for $n(A \cup B \cup C) = 33$.

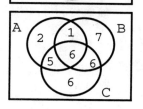

17. $460 = 240 + n(B) - 55$, $n(B) = 275$, so $n(B') = 500 - 275 = 225$

19.   We can show the given information in a Venn diagram. From it we
       have
       (a) 14   (b)   24
       (c) 14 + 24 = 38
       (d) 14 + 6 + 24 = 44

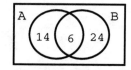

21.   Here is a Venn diagram of the
       given information.
       (a) 8   (b) 27
       (c) 4   (d) 31

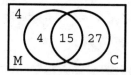

23.   Here is a Venn diagram of the given information.
       (a) 21   (b) 10
       (c) 8   (d) 56
       (e) 29   (f) 62

24.

25.   Here is a Venn diagram of the given information

       (a) 28   (b) 9
       (c) 20   (d) 7

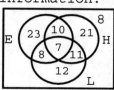

27. Here is a Venn diagram of the given information.

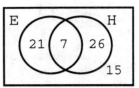

A total of 69 were present.

29. It must be the case that $n(A \cup B) \geq n(A)$, but it is not.

31. We show the given information in a Venn diagram.

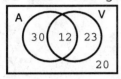

(a) 30   (b) 20

33. We show the given information in a Venn diagram.

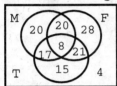

The information is inconsistent since these numbers total only 133, not 135.

35. The information reported a total of 56 people and they can be represented in the Venn diagram like this.

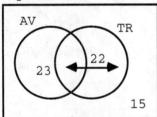

$n(TR \cup AG) = 22 + 23 = 45$
Total number of people = 45 + 15 = 60 which is not consistent with the reported 56 total.

37. (a) $15 = n(A \cup B) = n(A) + n(B) - n(A \cap B) = 15 - n(A \cap B)$,
    $n(A \cap B) = 0$
    (b) $A \cap B$) = empty set.

39. When $A \subset B$ the Venn diagram is

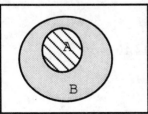

$A \cup B$ is the area occupied by B.

Section 5.3

1.

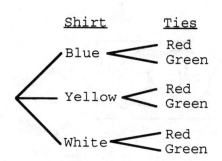

3.

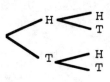

5. $12 \cdot 15 = 180$ 7. $5 \cdot 9 = 45$ 9. $6 \cdot 7 = 42$

11. $4 \cdot 7 = 28$

13. (a) $13 \cdot 13 = 169$ (b) $13 \cdot 13 \cdot 13 \cdot 13 = 28,561$

15. $2 \cdot 4 \cdot 3 = 24$ 17. $6 \cdot 5 \cdot 3 = 90$

19. Since Reshma buys just one item, there are $13 + 9 = 22$ choices.

21. $4^6 = 4,096$ 23. $8 \cdot 10^6 = 8,000,000$

25. (a) $6^4 = 1296$ (b) $6 \cdot 5 \cdot 4 \cdot 3 = 360$

27. (a) $26^3 \cdot 10^3 = 17,576,000$ (b) $26 \cdot 25 \cdot 24 \cdot 10 \cdot 9 \cdot 8 = 11,232,000$

29.

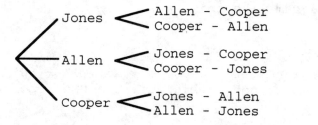

31. (a) $7 \cdot 6 \cdot 5 = 210$ (b) $1 \cdot 6 \cdot 5 = 30$ (c) $2 \cdot 6 \cdot 5 = 60$
     (d) $6 \cdot 2 \cdot 5 = 60$ (e) $2 \cdot 5 \cdot 4 + 5 \cdot 2 \cdot 4 + 5 \cdot 4 \cdot 2 = 3 \cdot 40 = 120$

33. $8 \cdot 1 \cdot 6 \cdot 1 \cdot 4 \cdot 1 \cdot 2 \cdot 1 = 384$

35. Men must sit in the first and seventh seat so $4 \cdot 3 \cdot 3 \cdot 2 \cdot 2 \cdot 1 \cdot 1 = 144$

37. (a) $6 \cdot 5 \cdot 4 \cdot 3 = 360$ (b) $6 \cdot 5 \cdot 5 \cdot 5 = 750$

39. $5 \cdot 4 \cdot 3 \cdot 2 \cdot 1 \cdot 3 \cdot 2 \cdot 1 + 3 \cdot 2 \cdot 1 \cdot 5 \cdot 4 \cdot 3 \cdot 2 \cdot 1 = 2 \cdot 720 = 1440$

41. $2^7 = 128$

43.

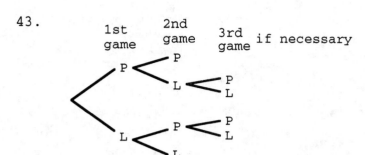

45. There are 26 × 26 × 26 = 17,576 different sets of three initials so it is not possible to index all slides with three initials. If four initials are used, there are 26^4 = 456,976 possibilities which is more than adequate for the 22,500 slides.

47. (a) Since there are 10 choices for each digit, there are 10^9 = 1,000,000,000 possible social security numbers.

(b) There are 26 choices for each of the six characters so there are 26^6 = 308,915,776 possible social security numbers, less than the current system.

(c) If all ten digits and all 26 letters are used, there are 36^6 = 2,176,782,336 possible social security numbers possible. This provides more possibilities than the current system.

If the 26 letters of the alphabet and the 8 digits 2-9 are used, there are 34^6 = 1,544,804,416 possibilities. This system will also provide a greater number of possibilities than the current system.

49. We need to find the number of six-character, seven-character, and eight-character passwords and add the results.

Six-character	26^6 =	308,915,776
Seven-character	26^7 =	8,031,810,176
Eight-character	26^8 =	$2.088270646 \times 10^{11}$
Total		2.1716×10^{11}

which is about 217 billion.

Section 5.4

1. $3! = 6$

3. $5! = 120$

5. $5!3! = 120(6) = 720$

7. $\dfrac{7!}{3!} = \dfrac{5040}{6} = 840$

9. $\dfrac{12!}{7!} = 12(11)(10)(9)(8) = 95,040$

11. $P(6, 4) = 6 \cdot 5 \cdot 4 \cdot 3 = 360$

13. $P(100, 3) = 100 \cdot 99 \cdot 98 = 970,200$

15. $P(7, 4) = 7 \cdot 6 \cdot 5 \cdot 4 = 840$

17. $P(6, 3) = 6 \cdot 5 \cdot 4 = 120$

19. $P(7, 5) = 7 \cdot 6 \cdot 5 \cdot 4 \cdot 3 = 2,520$

21. $P(8, 3) = 8 \cdot 7 \cdot 6 = 336$

23. (a) $P(7, 7) = 7! = 5,040$ (b) $P(7, 3) = 7 \cdot 6 \cdot 5 = 210$

25. $P(3, 3) = 3! = 6$

27. $P(22, 3) = 22 \cdot 21 \cdot 20 = 9,240$

29.   (a)   P(4, 4) = 4! = 24   (b)   P(7, 4) = 7·6·5·4 = 840

31.   7·6·5 = 210   33.   P(5, 5) = 5! = 120

35.   P(4, 4) = 4! = 24

37.   (a)   P(4, 3) = 4·3·2 = 24   (b) 4^3 = 64

39.   P(12, 2)·P(11, 5)·P(10, 2) = (12·11)(11·10·9·8·7)(10·9)
        = 132(55440)(90) = 658,627,200

41. $\frac{8!}{2!}$ = 20,160   43. $\frac{6!}{3!}$ = 120

45.   (a) $\frac{8!}{2!2!}$ = 10,080   (b) $\frac{11!}{4!4!2!}$ = 34,650

       (c) $\frac{10!}{2!2!3!}$ = 151,200   (d) $\frac{10!}{3!2!2!}$ = 151,200

47.   (a) $\frac{7!}{4!3!}$ = 35   (b) $\frac{7!}{5!2!}$ = 21

49.   (a)   The series ends when team A wins 4 games so the number of
              games played may range from 4 to 7. However, assume that A
              wins 4 and B wins 3 games (B "wins" any unplayed games). The
              number of sequences of games possible if A wins the series
              is the number of sequences of 4 wins by A and 3 wins by B,

              $\frac{7!}{4!3!}$ = 35

       (b)   35   (c)   35 + 35 = 70

51.   P(10, 4) = 10·9·8·7 = 5,040

53.   P(28, 3) = 28·27·26 = 19,656

55.   (a)   P(n, 2) = n(n - 1)   (b)   P(n, 3) = n(n - 1)(n - 2)
       (c)   P(n, 1) = n
       (d)   P(n, 5) = n(n - 1)(n - 2)(n - 3)(n - 4)

57.   26·9·26·9·26 = 1,423,656

59.   P(15, 4) = 15·14·13·12 = 32,760

61.   (a) $\frac{7!}{4!3!}$ = 35   (b) $\frac{8!}{5!3!}$ = 56

63.   P(n, k) is the product of the first k factors of the product
              n(n - 1)(n - 2)(n - 3) ... 3·2·1 = n!
       Thus, if we divided out the last n - k factors from n! we have
       P(n, k). The last n - k factors of n! are the numbers 1 through
       n - k so their product is (n - k)!
       Thus, $\frac{n!}{(n - k)!}$ leaves the first k factors of n! which is P(n, k).

69.   (a)   The cards can be shuffled in 52! = 8.0658 × 10^{67}.
       (b)   The volume occupied by one deck of cards is 2 × 3 × .5 = 3
              cubic inches.
              Total volume = 3 × 8.0658 × 10^{67} cubic inches
                           = 24.1974 × 10^{67} cubic inches

Section 5.5

Divide total volume by $12 \times 12 \times 12 = 1728$ to obtain cubic feet.

Total volume $= \dfrac{24.1974 \times 10^{67}}{1728} = 1.4003 \times 10^{65}$ cubic feet.

Divide by $5280^3 = 1.4719 \times 10^{11}$ to convert to cubic miles

$\dfrac{1.4003 \times 10^{65}}{1.4719 \times 10^{11}} = 9.5136 \times 10^{53}$ cubic miles

Length of one side of the cube $= \sqrt[3]{9.5131 \times 10^{53}}$ miles
$= 9.8530 \times 10^{17}$ miles

Since light travels 186,000 miles per second, the length of one side is $\dfrac{9.8352 \times 10^{17}}{186,000}$ light seconds.

$= 5.2877 \times 10^{12}$ light seconds.

We change this to light years by dividing by the number of seconds in a year, $60 \times 60 \times 24 \times 365 = 31,536,000$

$\dfrac{5,2877 \times 10^{12}}{31,536,000} = 1.6767 \times 10^{5}$ light years.

The box that will hold 52! decks of cards is a box with length, width, and height each 167,670 light years long.

Section 5.5

1. $C(6, 2) = \dfrac{6 \cdot 5}{2} = 15$ 3. $C(13, 3) = \dfrac{13 \cdot 12 \cdot 11}{3 \cdot 2} = 286$

5. $C(9, 5) = \dfrac{9 \cdot 8 \cdot 7 \cdot 6 \cdot 5}{5 \cdot 4 \cdot 3 \cdot 2 \cdot 1} = 126$ 7. $C(4, 4) = \dfrac{4!}{4!} = 1$

9. {a, b}, {a, c}, {a, d}, {b, c}, {b, d}, {c, d}

11. {a, b, c, d}, {a, b, c, e}, {a, b, d, e}, {a, c, d, e}, {b, c, d, e}

13. $C(6, 2) = \dfrac{6 \cdot 5}{2 \cdot 1} = 15$ 15. $C(15, 3) = \dfrac{15 \cdot 14 \cdot 13}{3 \cdot 2} = 45$

17. (a) $C(7, 4) = \dfrac{7 \cdot 6 \cdot 5 \cdot 4}{4 \cdot 3 \cdot 2} = 35$ (b) $P(7, 4) = 7 \cdot 6 \cdot 5 \cdot 4 = 840$

19. $C(5, 2) \cdot C(6, 3) \cdot C(8, 2) = (10)(20)(28) = 5600$

21. $C(6, 3) \cdot C(10, 4) = (20)(210) = 4200$

23. $C(9, 3) + C(11, 3) = 84 + 165 = 249$

25. $C(7, 3) + C(6, 3) = 35 + 20 = 55$

27. At least three means that 3, 4, or 5 men are on the committee.
$C(5, 3) \cdot C(6, 2) + C(5, 4) \cdot C(6, 1) + C(5, 5) = 10 \cdot 15 + 5 \cdot 6 + 1 = 181$

29. At least two American means 2, 3, or 4 American.
$C(8, 2) \cdot C(6, 2) + C(8, 3) \cdot C(6, 1) + C(8, 4) = 28 \cdot 15 + 56 \cdot 6 + 70 = 826$

31. $P(40, 2) \cdot C(10, 3) = 1560 \cdot 120 = 187,200$

33. $P(8, 2) \cdot C(12, 4) = 56 \cdot 495 = 27,720$

35. $C(7, 3) = 35$ 37. $C(20, 5) = 15,504$

39. (a) {Alice}, {Bianca}, {Cal}, {Dewayne}
    (b) {Alice, Bianca}, {Alice, Cal}, {Alice, Dewayne},
        {Bianca, Cal}, {Bianca, Dewayne}, {Cal, Dewayne}
    (c) {Alice, Bianca, Cal}, {Alice, Bianca, Dewayne},
        {Bianca, Cal, Dewayne}, {Alice, Cal, Dewayne},
    (d) {Alice, Bianca, Cal, Dewayne}   (e) Ø   (f) 16

41. $C(4, 2) \cdot C(4, 3) = 6 \cdot 4 = 24$

43. (a) $C(10, 6) \cdot C(8, 5) = (210)(56) = 11,760$
    (b) $C(8, 6) \cdot C(10, 4) + C(8, 7) \cdot C(10, 3) + C(8, 8) \cdot C(10, 2)$
        $= 5880 + 960 + 45 = 6885$

47. (a) Since the order of selection is irrelevant, the six numbers
        can be chosen in $C(50, 6) = 15,890,700$ different ways.
    (b) Measure a small stack of pennies. I found 17 pennies makes
        a stack one inch tall. Thus, a stack of 15,890,700 pennies
        makes a stack
        $$\frac{15,890,700}{17} = 934,747 \text{ inches tall.}$$
        $$\text{Height in feet} = \frac{934,747}{12} = 77,896 \text{ feet.}$$
        $$\text{Height in miles} = \frac{77,896}{5280} = 14.753 \text{ miles.}$$

51. There are 26 cards that are not red (all black) so there are
    $C(26, 5) = 65,780$ such hands.

55. The number is $C(52, 13) = 6.35 \times 10^{11}$.

57. (a) The number is $P(45, 6) = 5,864,443,200$.
    (b) The number is $C(45, 6) = 8,145,060$.

Section 5.6

1. This is an addition rule problem with the number of possible
   selections $= 21 + 23 = 44$.

3. A combination problem with $C(8, 3) = 56$ ways the three can be
   selected.

5. This is a multiplication rule with one item selected from each of
   3 sets. The number of ways the selection can be made is $56 \times 8 \times 15 = 6720$ ways.

7. This is a two level problem with the first level being a
   multiplication rule problem since selections are made from each of
   two sets. The selection from each set is a permutation problem
   since three different positions are being filled in each class.
        Number of selections of the student council
        $= P(110, 3) \times P(90, 3) =$
        $= 1,294,920 \times 704,880$
   which is about 9.128×10^{11}.

9. $C(11, 2) + 23 = 78$

11. $C(10, 3) = 120$

13. $5 \times 8 \times 3 = 120$

15. $8! = 40,320$

17. $n(\text{Barr} \cup \text{Sendon}) = 28 + 21 - 4 = 45$

19.   Select three from the 15 to go to Atlanta and select 4 from the remaining 12 to go to San Diego
      $C(15, 3) \times C(12, 4)$ $455 \times 495 = 225,225$

Section 5.7

1. $\dfrac{12!}{(3!)^4} = 369,600$

3. $\dfrac{7!}{3!4!} = 35$

5. $\dfrac{9!}{2!3!4!} = 1260$

7. $\dfrac{6!}{2!4!} = 15$

9. $\dfrac{9!}{(3!)^3} = 1680$

11. $\dfrac{14!}{3!5!6!} = 168,168$

13. $\dfrac{18!}{(6!)^3} = 17,153,136$

15. $\dfrac{15!}{6!4!5!} = 630,630$

17. $\dfrac{10!}{4!4!2!} = 3,150$

19. $\dfrac{10!}{(5!)^2} = 252$

21. $\dfrac{18!}{(6!)^3} = 17,153,136$

23. $\dfrac{9!}{(3!)^4} = 280$

25. $\dfrac{12!}{3!(4!)^3} = 5775$

27. $\dfrac{15!}{2!(5!)^2 3!2!} = 3,783,780$

29. $\dfrac{22!}{3!(2!)^3 4!(4!)^4}$

31. $\dfrac{50!}{10!(5!)^{10}}$

Review Exercises, Chapter 5

1.   (a) True   (b) False   (c) False   (d) False
     (e) True   (f) False   (g) True   (h) False
     (i) True   (j) False   (k) True   (l) False
     (m) True   (n) False   (o) False

3.   (a) Equal   (b) Not equal   (c) Equal, both are empty

5. $n(A \cap B) = 32 + 40 - 58 = 14$

7.

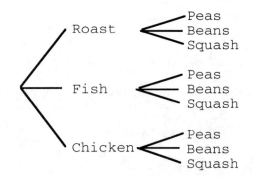

Roast — Peas, Beans, Squash
Fish — Peas, Beans, Squash
Chicken — Peas, Beans, Squash

9.   (a) $10^4 \cdot 26^2 = 6{,}760{,}000$ (b) $10 \cdot 9 \cdot 8 \cdot 7 \cdot 26 \cdot 25 = 3{,}276{,}000$

11. $C(15, 5) = 3{,}003$ 13. $(40)(27)(85)(34) = 3{,}121{,}200$

15. $4 \cdot 5! \cdot 3 = 1{,}440$

17.   (a) $P(11, 3) = 990$ (b) $11^3 = 1{,}331$

19. $P(10, 2) \cdot C(8, 4) = (90)(70) = 6300$

21.   The Venn diagram of the information is

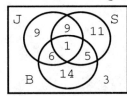

J, S, B
9 9 11
1
6 5
14 3

The total of persons is 58, not 60.

23.   Here is the Venn diagram of the given information.

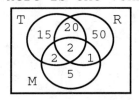

T, R, M
15 20 50
2
2 1
5

(a)   20   (b)   6   (c)   23

25. $5 \cdot 8 \cdot 7 = 280$ 27. $C(5, 2) = 10$

29.   (a) $(20)^3 = 8{,}000$ (b) $20 \cdot 19 \cdot 18 = 6{,}840$

31. $C(15, 4)C(20, 4)C(25, 3)C(11, 1)$

33. $P(9, 5) = 15{,}120$ 35. $C(22, 5) = 26{,}334$

37.   (a) $P(8, 4) = 8 \cdot 7 \cdot 6 \cdot 5 = 1{,}680$ (b) $C(9, 5) = \dfrac{9 \cdot 8 \cdot 7 \cdot 6 \cdot 5}{6 \cdot 5 \cdot 4 \cdot 3} = 126$

(c) $P(7, 7) = 7! = 5{,}040$ (d) $C(5, 5) = 1$

(e) $4! = 24$ (f) $\dfrac{7!}{3!4!} = 35$ (g) $\dfrac{8!}{4!} = 1{,}680$

39. $P(20, 2)P(15, 1)P(25, 1) = (380)(15)(25) = 142{,}500$

41. $83 = 46 + 51 - n(A \cap B)$
$n(A \cap B) = 14$ Fourteen belong to both.

43. $C(10, 4) = 210$

45. $\emptyset$, {red}, {white}, {blue}, {red, white}, {red, blue}, {white, blue}, {red, white, blue}

47. $P(10, 2) \cdot C(8, 3) = (90)(56) = 5,040$

49. $C(6, 4) + C(8, 4) = 15 + 70 = 85$

51. (a) $5! = 120$ (b) $\dfrac{5!}{3!} = 20$ (c) $\dfrac{8!}{2!2!} = 10,080$

53. $\dfrac{18!}{2!(3!)^2 2!(6!)^2} = 85,765,680$

55. $\dfrac{12!}{2!(3!)^2(2!)(3!)^2} = 92,400$

Chapter 6
Probability

Section 6.1

1. (a) {True, False}
   (b) {vowel, consonant} or the letters {a,b,c, . . . ,x,y,z}
   (c) {1, 2, 3, 4, 5, 6}
   (d) {HHH, HHT, HTH, THH, TTH, THT, HTT, TTT}
   (e) {Mon, Tue, Wed, Thurs, Fri, Sat, Sun}
   (f) {Grand Canyon wins, Bosque wins, tie}, or
       {Bosque wins, Bosque loses, tie}
   (g) {pass, fail}, {A, B, C, D, F}
   (h) {Susan & Leah, Susan & Dana, Susan & Julie, Leah &
          Dana, Leah & Julie, Dana and Julie}
   (i) {Mon, Tue, Wed, Thur, Fri, Sat, Sun}, or {1, 2, 3, 4,...,31}
   (j) {male, female}, {passing student, failing student}

3. (a) Valid   (b) Valid
   (c) Invalid, $P(T) + P(D) + P(H) = .95$
   (d) Invalid $P(2) + P(4) + P(6) + P(8) + P(10) = 1.15$
   (e) Valid   (f) Invalid $P(Utah) = -.4 < 0$
   (g) Valid   (h) Invalid $P(maybe) = 1.1 > 1$
   (i) Invalid $P(True) + P(False) = 0$

5. (a) $P(\{A, B\}) = .1 + .2 = .3$
   (b) $P(\{B, D\}) = .2 + .4 = .6$
   (c) $P(\{A, C, D\}) = .1 + .3 + .4 = .8$
   (d) $P(\{A, B, C, D\}) = 1$

7. (a) $.10 + .37 = .47$ (b) $.05 + .18 + .30 = .53$

9. (a) $.05 + .10 + .06 = .21$ (b) $.10 + .15 + .18 = .43$
   (c) $.21 + .11 + .18 + .14 = .64$ (d) $.12 + .15 = .27$

11. (a) The table of relative frequencies is obtained by dividing
        each entry by 81.

	Ms. Busby	Mr. Butler	Mrs. Hutchison
Calculator	.160	.136	.210
No Calculator	.185	.198	.111

    (b) P(Calculator) = .160 + .136 + .210 = .506
    (c) P(Ms. Busby's student and no calculator) = .185
    (d) P(Mr. Butler) = .136 + .198 = .334

13. $P(A) + P(B) + P(C) + P(D) = 2 \cdot P(D) + 3 \cdot P(D) + 4 \cdot P(D) + P(D) = 1$
    $10 \cdot P(D) = 1$ so $P(D) = .1$
    $P(A) = .20, P(B) = .30, P(C) = .40, P(D) = .10$

15. $P(A) + P(B) + P(C) = 1$
    $P(A) + P(B) \qquad\quad = .75$
    $\qquad\quad P(B) + P(C) = .45$
    Subtract second equation from the first
    $\qquad\quad P(A) + P(B) + P(C) = 1$
    $\qquad\quad \underline{P(A) + P(B) \qquad\quad = .75}$
    $\qquad\qquad\qquad\qquad P(C) = .25$
    From third equation $P(B) + .25 = .45$ gives $P(B) = .20$, so $P(A) = .55$.

17. P(Mini) = 140/800 = .175, P(Burger) = 345/800 = .431,
     P(Big B) = 315/800 = .394

19. P(below 40) = 160/1800 = .089,
     P(40-49) = 270/1800 = .15
     P(50-55) = 1025/1800 = .569,
     P(over 55) = 345/1800 = .192

21. (a) P({2, 4}) = .16 + .30 = .46
     (b) P({1, 3, 5}) = .15 + .19 + .20 = .54
     (c) P(prime) = .16 + .19 + .20 = .55

23. (a) 4/52 = 1/13   (b) (4 + 4)/52 = 2/13
     (c) 13/52 = 1/4   (d) 26/52 = 1/2

Section 6.2

1. 4/15

3. n = the total number of sweaters, 36. s = the number of ways to
    select one small size, 22.
    $P(small) = \dfrac{22}{36} = \dfrac{11}{18} = .611$.

5. There are 6(6) = 36 possible ways for two dice to turn up. Of
    these, a seven occurs in 5 ways so p = 6/36 = 1/6

7. $\dfrac{C(10,\ 2)}{C(17,\ 2)} = \dfrac{45}{136} = .33$

9. $\dfrac{C(4,\ 2)}{C(52,\ 2)} = \dfrac{6}{1326} = \dfrac{1}{221} = .0045$

11. (a) $P(stamped) = \dfrac{10}{18} = \dfrac{5}{9}$ (b) $P(not\ stamped) = \dfrac{8}{18} = \dfrac{4}{9}$

    (c) $P(not\ stamped\ or\ stamped) = \dfrac{18}{18} = 1$

13. Two tosses can occur in 4 ways. Except for the one case when both
    are tails, at least one is a head. Thus, p = 3/4

15. $\dfrac{C(12,\ 2)\cdot C(9,\ 1)}{C(21,\ 3)} = \dfrac{594}{1330} = \dfrac{297}{665} = .447$

17. (a) $\dfrac{C(4,\ 4)}{C(15,4)} = \dfrac{1}{1365} = .00073$

    (b) $\dfrac{C(4,\ 3)\cdot C(11,\ 1)}{C(15,\ 4)} = \dfrac{4\cdot 11}{1365} = \dfrac{44}{1365} = .0322$

    (c) $\dfrac{C(4,\ 2)\cdot C(11,\ 2)}{C(15,\ 4)} = \dfrac{6\cdot 55}{1365} = \dfrac{22}{91} = .242$

    (d) $\dfrac{C(11,\ 4)}{C(15,\ 4)} = \dfrac{330}{1365} = \dfrac{22}{91} = .242$

19. (a) $\dfrac{P(10,\ 2)\cdot P(6,\ 3)}{P(16,\ 5)} = \dfrac{90\cdot 120}{524160} = \dfrac{90}{4368} = \dfrac{15}{728} = .0206$

    (b) $\dfrac{10\cdot 6\cdot 9\cdot 5\cdot 8}{P(16,\ 5)} = \dfrac{10\cdot 6\cdot 9\cdot 5\cdot 8}{524160} = \dfrac{15}{364} = .0412$

21.   (a) $\dfrac{P(4,\ 4)}{P(8,\ 4)} = \dfrac{24}{1680} = \dfrac{1}{70} = .0143$

      (b) $\dfrac{P(4,\ 2) \cdot P(4,\ 2)}{P(8,\ 4)} = \dfrac{12 \cdot 12}{1680} = \dfrac{3}{35} = .0857$

23. $1 - (.6 + .3) = .1$

25.   Begins with a 5, 6, 7, 8 or 9: $\dfrac{5 \cdot 8 \cdot 7}{P(9,\ 3)} = \dfrac{280}{504} = \dfrac{5}{9} = .555$

27.   (a) $\dfrac{1}{P(5,\ 5)} = \dfrac{1}{5!} = \dfrac{1}{120}$ (b) $\dfrac{P(2,\ 2) \cdot P(3,\ 3)}{P(5,\ 5)} = \dfrac{2 \cdot 6}{120} = \dfrac{1}{10}$

29. $\dfrac{C(8,\ 3) \cdot C(6,\ 1)}{C(14,\ 4)} = \dfrac{56 \cdot 6}{1001} = \dfrac{48}{143} = .336$

31.   (a) $6/28 = 3/14$ (b) $3/28$ (c) $10/28 = 5/14$

33.   (a) $5/14$ (b) $(6 + 3)/14 = 9/14$

35.   (a) $400/500 = 4/5$ (b) $100/500 = 1/5$

37.   (a) $\dfrac{P(8,\ 3)}{P(15,\ 3)} = \dfrac{8 \cdot 7 \cdot 6}{15 \cdot 14 \cdot 13} = \dfrac{8}{65} = .123$

      (b) $\dfrac{8 \cdot 7 \cdot 7}{P(15,\ 3)} = \dfrac{8 \cdot 7 \cdot 7}{15 \cdot 14 \cdot 13} = \dfrac{28}{195} = .144$

      (c) $\dfrac{5 \cdot 4 \cdot 3}{P(15,\ 3)} = \dfrac{5 \cdot 4 \cdot 3}{15 \cdot 14 \cdot 13} = \dfrac{2}{91} = .022$

39. $324/570 = 54/95 = .568$

41.   Five diamonds can be selected in $C(13,5)$ and a selection of any five cards can occur in $C(52,\ 5)$ ways. Thus,

$$P(5 \text{ diamonds}) = \frac{C(13,5)}{C(52,5)} = \frac{1287}{2598960} = .000495$$

43.   The dice can turn up in 6^6 different ways.

     If all turn up a different number, then the first die will turn up some number, six possible ways. Then the second die can turn up any of the five remaining numbers, the third any of the remaining four numbers, and so on.
The six different numbers then can occur in
    $6 \times 5 \times 4 \times 3 \times 2 \times 1 = 720$ ways.

The probability of six different numbers is $p = \dfrac{720}{6^6} = .0154$.

45.   (a)   The draw can occur in $C(40,\ 6) = 3{,}838{,}380$ ways.
      (b)   The probability of winning is $1/3838380$.

47.   Let x = number of dimes dated before 1960
$p = x/250 = .2$ so $x = .2(250) = 50$

49. $\dfrac{C(2,\ 2)}{C(6,\ 2)} = \dfrac{1}{15}$

51.   (a) $\dfrac{3 \cdot 5 \cdot 2}{P(10,\ 3)} = \dfrac{3 \cdot 5 \cdot 2}{10 \cdot 9 \cdot 8} = \dfrac{1}{24}$ (b) $\dfrac{5 \cdot 4 \cdot 2}{P(10,\ 3)} = \dfrac{5 \cdot 4 \cdot 2}{10 \cdot 9 \cdot 8} = \dfrac{1}{18}$

      (c) $\dfrac{3 \cdot 2 \cdot 2}{P(10,\ 3)} = \dfrac{3 \cdot 2 \cdot 2}{10 \cdot 9 \cdot 8} = \dfrac{1}{60}$

53. (a) To win, the participant must select the six numbers drawn.
This can be done in C(6, 6) = 1 way.
The six numbers can be drawn in C(50, 6) = 15,890,700 ways

$$P(\text{Win first prize}) = \frac{1}{15,890,700} = .0000000629$$

(b) To win second prize, the participant must select 5 of the
six numbers drawn and select one of the 44 not drawn. This
can be done in C(6, 5) × C(44, 1) = 6 × 44 = 264 ways.

$$P(\text{second prize}) = \frac{264}{15,890,700} = .0000166$$

Section 6.3

1. E ∪ F = {1, 2, 3, 4, 5, 7, 9}   E ∩ F = {1, 3}
E' = {2, 4, 6, 8, 10}

3. E ∪ F = the set of students who are passing English or failing
chemistry (or both)
E ∩ F = the set of students who are passing English and failing
chemistry
E' = the set of students who are failing English

5. 1 − 3/5 = 2/5   7. P(E) = 1 − 0.7 = 0.3

9. 1 − 0.3 = 0.7   11. Mutually exclusive

13. Not mutually exclusive because a student can be taking both.

15. Mutually exclusive since there is no February 30.

17. (a) $\frac{C(4, 4)}{C(52, 4)} = \frac{1}{270725}$ (b) $\frac{1}{270725} + \frac{1}{270725} = \frac{2}{270725}$

19. 1/10 + 3/10 = 4/10 = 2/5

21. (a) Find 1 − probability all females.
$$1 - \frac{C(5, 3)}{C(12, 3)} = 1 - \frac{10}{220} = 1 - \frac{1}{22} = \frac{21}{22}$$

(b) Find 1 − probability all males.
$$1 - \frac{C(7, 3)}{C(12, 3)} = 1 - \frac{35}{220} = 1 - \frac{7}{44} = \frac{37}{44}$$

23. $1 - \frac{C(46, 3)}{C(50, 3)} = 1 - \frac{15180}{19600} = \frac{4420}{19600} = \frac{221}{980} = 0.226$ (1 − probability
none of 4 wins)

25. (a) 120/400 = 3/10   (b) 220/400 = 11/20   (c) 55/400 = 11/80

(d) $\frac{120 + 220 - 55}{400} = \frac{285}{400} = \frac{57}{80} = .713$

(e) 220 − 55 = 165 are in enrolled in English but not in
mathematics so p = $\frac{165}{400}$ = .413

(f) 500 − 220 = 280 are not enrolled in English,
120 − 55 = 65 are enrolled in mathematics and not in English
so n(not in English or in math) = 280 + 120 − 65 = 235 so p
= $\frac{235}{400}$ = .588

27. (a) 4/35 (b) $\dfrac{10 + 14 - 4}{35} = \dfrac{20}{35} = \dfrac{4}{7}$

29. .75 + .24 - .18 = 0.81   31. .36 + .26 - .11 = .51

33. (a) .85 + .60 - .55 = .90   (b) 1 - .90 = .10

35. (a) .75 + .63 - .54 = 0.84   (b) 1 - .84 = 0.16

37. (a) 3/6 = 1/2   (b) 2/6 = 1/3
    (c) 3/6 = 1/2   (d) 4/6 = 2/3

39. Mutually exclusive   41. Mutually exclusive

43. (a) 5/6   (b) 1/6   (c) 3/6 = 1/2
    (d) 2/6 = 1/3   (e) 2/6 = 1/3

45. (a) 1/10   (b) 1/10   (c) 5/10 = 1/2
    (d) 5/10 = 1/2   (e) 5/10 = 1/2
    (f) (5 + 6 - 3)/10 = 8/10 = 4/5
    (g) (5 + 7 - 3)/10 = 9/10

47. (4 + 13 - 1)/52 = 16/52 = 4/13

49. .89 = .76 + .62 - P(passing both)   P(passing both) = .49

51. P(Female or Executive)
    = P(Female) + P(Executives) - P(Females and Executives)
    .65 = .55 + P(Executives) - .05, so
    P (Executives) = .15
    Since .05 are female executives, the remainder
    .10 = 10%, are male executives.

53. (3 + 1)/6 = 4/6 = 2/3

55. (a) P(E) = 4/10 = 2/5 = probability he attends a small or large
        state university
        P(F) = 6/10 = 3/5 = probability he attends a small or large
        private university
        P(G) = 5/10 = 1/2 = probability he attends a large
        university
        P(H) = 5/10 = 1/2 = probability he attends a small state
        university or a small private university
    (b) P(E') = 1 - 2/5 = 3/5 = probability he does not attend a
        state university
    (c) P(E $\cup$ G) = (3 + 2 + 1)/10 = 6/10 = 3/5 = probability he
        attends a large or state university
    (d) P(F' $\cap$ H) = 1/10 = probability he attends a small state
        university

57. P(At least one has the part) = .85 + .93 - .81 = .97

59. (a) Probability of at least one 6 =
        1 - probability of no sixes = $1 - \left(\dfrac{5}{6}\right)^{4}$ = .518
    (b) Probability of a pair of sixes =
        1 - probability of no pair of sixes = $1 - \left(\dfrac{35}{36}\right)^{26}$ = .519.
        They are essentially equally likely.

1.   (a)   3/4   (b)   1/2

3.   (a)   16/45   (b)   P(correct|from Section 1) = 7/24
      (c)   P(Section 2|correct) = 9/16

5.   (a)   3/8   (b)   7/10

7.   P(K|F) = 1/10, P(F|K) = 1

9.   (a)   28/60 = 7/15   (b)   6/20 = 3/10   (c)   6/28 = 3/14

11.   P(E|F) = .3/.7 = 3/7   P(F|E) = .3/.6 = 1/2

13.   P(E|F) = .24/.40 = 3/5   P(F|E) = .24/.60 = 2/5

15.   (a)   .20/.45 = 4/9 = .444   (b)   .20/.75 = 4/15 = .267

17.   (a)   41/497 = .0825   (b)   260/497 = .523
      (c)   145/497 = .292   (d)   46/260 = .177
      (e)   46/115 = .400
      (f)   (23 + 145)/(41 + 341) = 168/382 = .440

19.   (4/52)(4/51) = 4/663

21.   (a) $(4/52)(4/52) = (1/13)^2 = 1/169$

      (b) $(13/52)(13/52) = (1/4)^2 = 1/16$

      (c) $(26/52)(26/52) = (1/2)^2 = 1/4$

23.   (4/12)(2/11)(6/10) = 2/55

25.   (6/14)(5/13) = 15/91 = .165

27.   (3/32)(3/32) = 9/1024 = .0088

29.   (a)   (4/15)(4/15) = 16/225   (b)   (4/15)(3/15) = 4/75

31.   (a)   (6/15)(4/15)(6/15) = 16/375
      (b)   (5/15)(5/15)(5/15) = 1/27

33.   If one die shows a 2, the other die can show five different faces.
      If the second die shows a 2, the first one can show five different
      faces.   Thus, there are 10 possible ways one die shows a 2.   Of
      these 10 ways, two of them total 6. The probability of a total of
      6 given one die shows a 2 is 2/10.

35.   (a)   75/135 = 5/9   (b)   .35   (c)   .40
      (d)   Find P(Out of state and male or out of state and female) =
             P(O|M)·P(M) + P(O|F)·P(F) = (.35)(4/9) + (.40)(5/9)
                                       = .156 + .222 = .378

      (e) $\dfrac{P(F \cap O)}{P(O)} = \dfrac{P(F) \cdot P(O|F)}{P(O)} = \dfrac{(5/9)(.40)}{.378} = \dfrac{2/9}{.378} = .588$

37.   (a) $\left(\dfrac{2}{3}\right)\left(\dfrac{1}{2}\right) = \dfrac{1}{3}$ (b) $\left(\dfrac{2}{3}\right)\left(\dfrac{1}{2}\right) = \dfrac{1}{3}$

      (c) $\left(\dfrac{2}{3}\right)\left(\dfrac{1}{2}\right) + \left(\dfrac{1}{3}\right)\left(\dfrac{2}{2}\right) = \dfrac{2}{3}$

39. If one check in 10,000 is forged, 5% of all checks are P(forged) = 1/10,000   P(postdated) = .05
     P(postdated |forged) = .80

     Find P(forged |postdated) = $\dfrac{\text{P(forged and postdated)}}{\text{P(postdated)}}$

     Since P(postdated | forged) = $\dfrac{\text{P(forged and postdated)}}{\text{P(forged)}}$

     then .80 = $\dfrac{\text{P(forged and postdated)}}{.0001}$

     P(forged and postdated) = .00008

     P(forged | postdated) = $\dfrac{.00008}{.05}$ = .0016

41. P(scholarship and continues) =
         P(scholarship)P(continues | scholarship) = (.30)(.90) = .27

43. (a) $(\tfrac{1}{5})(\tfrac{1}{4})(\tfrac{1}{3}) = \dfrac{1}{60}$

     (b) $3!\dfrac{1}{60}$ = $\dfrac{1}{10}$ since 1, 2, and 3 can occur in 3! sequences.

           Or you can use $\dfrac{C(3, 3)}{C(5, 3)} = \dfrac{1}{10}$

     (c)   Not possible, p = 0

     (d)   There is only one set of numbers that add to 6, namely 1, 2,
           and 3 so p = $\dfrac{C(3, 3)}{C(5, 3)} = \dfrac{1}{10}$

     (e)   The numbers must be 1, 3, and 5 so p = $\dfrac{C(3, 3)}{C(5, 3)}$ = $\dfrac{1}{10}$

     (f) $\dfrac{C(3, 2)\ C(2, 1)}{C(5, 3)}$ = $\dfrac{3}{5}$

45. (a)   1/10   (b) $(\tfrac{9}{10})(\tfrac{8}{9}) = \dfrac{8}{10}$ = $\dfrac{4}{5}$

     (c)   Anita draws the unsolved problem and Al doesn't or Anita
           does not draw the unsolved problem and Al does not draw the
           unsolved problem, p = $\dfrac{1}{10}\dfrac{9}{9}$ + $\dfrac{9}{10}\dfrac{8}{9}$ = $\dfrac{9}{10}$

47. Let A represent those who passed the aptitude test and D represent
     those who passed the drug test.

     (a)   P(A ∩ D) = $\dfrac{42}{96}$ = .438

     (b)   P(A) = $\dfrac{54}{96}$ = .563

     (c)   P(D | A) = $\dfrac{\text{P(A ∩ D)}}{\text{P(A)}}$ = $\dfrac{42}{54}$ = .778

49. P(E|F) is the fraction of the area of F
     that is occupied by E ∩ F and is
     represented by the shaded area.

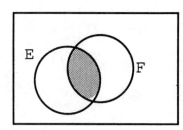

P(E'|F) is the fraction of the area of F
that is occupied by E' ∩ F and is
represented by the shaded area.

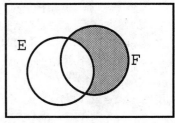

The area representing P(E|F) + P(E'|F) covers all of F so
P(E|F) + P(E'|F) = 1.

51. (a) A wager is good if the probability of winning is greater
        than one-half.

    (b) (i) The color of the first card is irrelevant. The bet is
            won if the second card is a different color.

        P(2 different colors) = $1 \times \dfrac{5}{9} = \dfrac{5}{9}$ This is a good wager.

        (ii) P(2 different numbers) = $1 \times \dfrac{8}{9} = \dfrac{8}{9}$

             This is a good wager.

    (c) (i) P(3 different colors) = $1 \times \dfrac{10}{14} \times \dfrac{5}{13}$ = .275

            This is not a good wager.

        (ii) P(3 different numbers) = $1 \times \dfrac{12}{14} \times \dfrac{9}{13}$ = .593

             This is a good wager.

    (d) (i) P(4 different colors) = $1 \times \dfrac{15}{19} \times \dfrac{10}{18} \times \dfrac{5}{17}$ = .129

            This is not a good wager.

        (ii) P(4 different numbers) = $1 \times \dfrac{16}{19} \times \dfrac{12}{18} \times \dfrac{8}{17}$ = .264

             This is not a good wager.

53.   The problem gives the following information.
       The sample space is the set of all registered voters
       n(Registered voters) = 84,349
       n(Democrats who voted) = 15,183
       n(Republicans who voted) = 16,026
       n(people who voted) = 15,183 + 16,026 = 31,209
       n(voted for Arnold) = 16,827
       n(voted for Betros) = 14,382
       P(voted for Arnold | Democrat) = .39
       P(voted for Betros | Republican) = .32

    (a) P(Registered and did not vote) = $\dfrac{84,349 - 31,209}{84,349}$

                                       = $\dfrac{53,140}{84,349}$ = .630

    (b) P(Democrat who voted) = $\dfrac{15,183}{84,349}$ = .180

    (c) P(voted for Betros) = $\dfrac{14,382}{84,349}$ = .171

    (d) P(voted for Arnold | voted) = $\dfrac{16,827}{31,209}$ = .539

(e) We want to find P(Republican | voted for Betros).
     First of all, we are restricted to those who voted, and we
     need n(Republican and voted for Betros), and n(voted for
     Betros).
     Those numbers are 5128 = .32 × 16,026 and 14,382,
     respectively.

     Thus, P(Republican|Voted for Betros) = $\frac{5128}{14382}$ = .357

Section 6.5

1. (a) (i)   Not mutually exclusive
         (ii) Independent, since P(E) = 40/160 = 1/4 and
               P(E|F) = 10/40 = 1/4
    (b) (i)   Not mutually exclusive
         (ii) Dependent, since P(E) = 40/160 = 1/4
               and P(E|F) = 10/50 = 1/5
    (c) (i)   Not mutually exclusive
         (ii) Independent, since P(E) = 50/100 = 1/2
               and P(E|F) = 30/60 = 1/2
    (d) (i)   Mutually exclusive

         (ii) Dependent, since P(E ∩ F) = 0 ≠ P(E)P(F) = $\frac{1}{16}$
    (e) (i)   Mutually exclusive
         (ii) Dependent, since P(E) = 30/120 = 1/4
               and P(E|F) = 0/40 = 0
    (f) (i)   Mutually exclusive
         (ii) Independent, since P(E) = 0/80 = 0
               and P(E|F) = 0/20 = 0

3. Independent, since (0.3)(0.5) = 0.15

5. Dependent, since (.3)(.7) ≠ .20

7. (a) Not mutually exclusive since one ring has both a diamond and
         a ruby.
    (b) Independent since P(E) = 2/4 = 1/2 and P(E|F) = 1/2

9. Dependent, since P(A ∩ S) = 10/100 = 1/10
    and P(A)·P(S) = (15/100)(20/100) = 3/100

11. (6/14)(12/21) = (3/7)(4/7) = 12/49

13. (1/4)(1/5) = 1/20

15. (a) (.2)(.3) = .06   (b) (.8)(.7) = .56
    (c) (.2)(.7) + (.8)(.3) = .14 + .24 = .38

17. (a) (.5)(.6)(.8) = .24   (b) (.5)(.4)(.2) = .04

19. (a) .3 + .5 - (.3)(.5) = .8 - .15 = .65
    (b) .2 + .6 - (.2)(.6) = .8 - .12 = .68
    (c) .4 + .6 - (.4)(.6) = 1 - .24 = .76

21. (a) Find 1 - P(neither shows up) = 1 - (.4)(.2) = 1 - .08 = .92
    (b) Find 1 - P(both show up) = 1 - (.6)(.8) = 1 - .48 = .52

23. 1 - (.7)(.1) = 1 - .07 = .93

Section 6.5

25. P(F) = 12/52 = 3/13, P(G) = 4/52 = 1/13,
P(G|F) = 4/12 = 1/3, P(F|G) = 1, dependent since P(G) ≠ P(G|F)
Or you can use P(F) = 3/13 ≠ P(F|G) = 1

27. (1/6)(1/6)(1/6)(3/6)(3/6) = 1/864

29. (a)

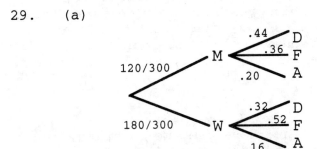

(b) Find the probability of each branch that ends in F and add. p = (120/300)(.36) + (180/300)(.52) = 0.456

(c) (120/300)(.44) = 0.176
(d) P(F) = 0.456, P(F|W) = .52, so dependent

31. P(F)·P(G) = (1/2)(3/13) = 3/26, P(F ∩ G) = 6/52 = 3/26, so independent

33. (a) $(.5)^{12}$ = .000244   (b) $(.90)^{12}$ = 0.282

35. This succeeds when the nickel turns up heads and the quarter turns up tails twice or the nickel and quarter turn up tails.
(3/5)(1/3)(1/3) + (2/5)(1/3) = 1/15 + 2/15 = 3/15 = 1/5

37. P(met verbal) = $\frac{2850}{3200}$ = .890

P(met verbal|met math) = $\frac{2795}{2965}$ = .943

These data indicate that meeting the verbal and meeting the math requirements are dependent since P(met verbal) ≠ P(met verbal | met math)

39. P(Female) = $\frac{135}{297}$ = .4545   P(Female|Science) = $\frac{40}{88}$ = .4545
The events female student and science major are independent.

41. Denote at most one head by H and at least one head and one tail by B. Let's compare P(H) and P(H|B).
The four tosses can turn up in 2^4 = 16 ways. At most one head occurs when all four are tails (1 way) or one head and 3 tails (4 ways). From this we have
P(H) = $\frac{5}{16}$

Of the 16 possible outcomes only the two cases of all heads or all tails fail to give at least one head and one tail so B can occur in 14 ways. Of the 14 ways we can obtain at least one head and one tail, how many have at most one head? Since all of these have at least one head, then at most one head occurs when there is exactly one head, four cases. Thus, P(H|B) = $\frac{4}{14}$.

Since P(H) ≠ P(H | B) the events are dependent.

43.   If E and F are mutually exclusive, then $P(E \cap F) = 0$. If $P(E)P(F) = 0$, then one of $P(E)$, $P(F)$ is zero. Thus, E and F (mutually exclusive) are independent when at least one of $P(E)$ or $P(F)$ is zero.
Example: Let the sample space be the set of United States citizens and let E = females who are currently president or vice-president of the United States. Let F = females who are currently members of the United States Senate. One person is selected at random.
Since E is empty, $p(E) = 0$. Since F is not empty, $P(F) \neq 0$.
However E and F are mutually exclusive so $P(E \cap F) = 0$.
Thus, $P(E)P(F) = 0 \times P(F) = 0 = P(E \cap F)$ so E and F are independent.

45.   A and C may or may not be independent. Here are two examples, the first A and C are not independent, and the second A and C are independent.
First, let A, B, and C be related as shown in the Venn diagram with the number of elements in each region shown.

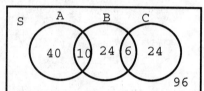

From the diagram we obtain the following
$n(S) = 200$
$n(A) = 50$
$n(B) = 40$
$n(C) = 30$
$n(A \cap B) = 10$
$n(B \cap C) = 6$
$n(A \cap C) = 0$

$P(A) = \dfrac{50}{200} \qquad P(B) = \dfrac{40}{200} \qquad P(C) = \dfrac{30}{200}$

$P(A \cap B) = \dfrac{10}{200} \qquad P(A \cap C) = 0 \qquad P(B \cap C) = \dfrac{6}{200}$

$P(A)\, P(B) = \dfrac{50}{200} \times \dfrac{40}{200} = \dfrac{20}{400} = \dfrac{10}{200} = P(A \cap B)$ so A and B are independent.

$P(B)P(C) = \dfrac{40}{200} \times \dfrac{30}{200} = \dfrac{12}{400} = \dfrac{6}{200} = P(B \cap C)$
so B and C are independent.

$P(A)P(C) = \dfrac{50}{200} \times \dfrac{30}{200} = \dfrac{15}{400} \neq P(A \cap C) = 0$
so A and C are not independent.
For the second example use the following Venn diagram.

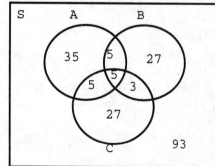

From this diagram we obtain
$n(S) = 200$
$n(A) = 50$
$n(B) = 40$
$n(C) = 40$
$n(A \cap B) = 10$
$n(A \cap C) = 10$
$n(B \cap C) = 8$

$$P(A) = \frac{50}{200} \qquad P(B) = \frac{40}{200} \qquad P(C) = \frac{40}{200}$$

$$P(A \cap B) = \frac{10}{200} \qquad P(A \cap C) = \frac{10}{200} \qquad P(B \cap C) = \frac{8}{200}$$

$$P(A)P(B) = \frac{50}{200} \times \frac{40}{200} = \frac{20}{400} = \frac{10}{200} = P(A \cap B)$$

so A and B are independent.

$$P(B)P(C) = \frac{40}{200} \times \frac{40}{200} = \frac{16}{400} = \frac{8}{200} = P(B \cap C)$$

so B and C are independent.

$$P(A)P(C) = \frac{50}{200} \times \frac{40}{200} = \frac{20}{400} = \frac{10}{200} = P(A \cap C)$$

so A and C are independent.

This example shows that it is possible for A and C to be independent when both A and B are independent and B and C are independent.

49. For Ron, the probability of at least one malfunction in a 30 day period is $1 - (.99)^{30} = .260$. Thus, Ron should not wait 30 days to back up his hard disk.
For Angel, the probability of at least one malfunction in a period of n days is $1 - (.9995)^n$. Find the largest n such that $1 - (.9995)^n < .01$. This is equivalent to finding $(.9995)^n > .99$. Trial and error shows that $n = 20$ is the largest value of n. Thus, Angel should backup her hard disk at least every 20 days.

Section 6.6

1.

3. (a) $P(E_1|F) = .7/.9 = 7/9 = .778$

(b) $P(E_1|F) = \dfrac{(.75)(.40)}{(.75)(.40) + (.25)(.10)}$

$$= \frac{.3}{.3 + .025} = 0.923$$

5. (a) $P(E_1 \cap F) = (5/12)(2/5) = 1/6,$
$P(E_2 \cap F) = (4/12)(1/4) = 1/12,$
$P(E_3 \cap F) = (3/12)(1/3) = 1/12$

(b) $P(F) = 1/6 + 1/12 + 1/12 = 1/3$

(c) $P(E_1|F) = \dfrac{\frac{1}{6}}{\frac{1}{3}} = \dfrac{1}{2} \qquad\qquad P(E_3|F) = \dfrac{\frac{1}{12}}{\frac{1}{3}} = \dfrac{1}{4}$

7. $P(II|D) = \dfrac{P(II)\ P(D|II)}{P(I)\ P(D|I) + P(II)\ P(D|II)} =$

$\dfrac{(.45)(.04)}{(.55)(.03) + (.45)(.04)} = 0.5217$

9. $P(H|W) = \dfrac{P(H)\ P(W|H)}{P(H)\ P(W|H) + P(Away)\ P(W|Away)} =$

$\dfrac{(.60)(.80)}{(.60)(.80) + (.40)(.55)} = 0.6857$

11. Use the notation So = sophomore, J = junior, Sr = senior, and A = received an A. The following information is given

$P(So) = \dfrac{10}{50} = .20,\quad P(J) = \dfrac{25}{50} = .50,\quad P(Sr) = \dfrac{15}{50} = .30,$

$P(A|So) = \dfrac{3}{10} = .30,\ P(A|J) = \dfrac{5}{25} = .20,\ P(A|Sr) = \dfrac{6}{15} = .40$

We are to find $P(J|A)$.

$P(J|A) = \dfrac{P(J)P(A|J)}{P(J)P(A|J) + P(So)P(A|So) + P(Sr)P(A|Sr)}$

$= \dfrac{.50(.20)}{.50(.20) + .20(.30) + .30(.40)}$

$= \dfrac{.1}{.1 + .06 + .12} = .357$

The probability a student who received an A is a junior is about .36.

13. (a)

(b) P(red|Box I) = 2/5
     P(red and Box I) = (3/5)(2/5) = 6/25
     P(red) = (3/5)(2/5) + (2/5)(5/12)
          = 6/25 + 1/6 = 61/150

(c) $P(Box\ I|red) = \dfrac{P(Box\ I\ and\ red)}{P(red)} = \dfrac{\frac{6}{25}}{\frac{61}{150}}$

$= \dfrac{6(150)}{61(25)} = 36/61$

15. Let A represent a driver involved in an accident, DE represent a driver with driver's ed, and no DE represent a driver with no DE. We are given
P(DE) = .80, P(no DE) = .20, P(A|DE) = .32, P(A|no DE) = .55
We are to find P(DE|A).

$P(DE|A) = \dfrac{P(DE)P(A|DE)}{P(DE)P(A|DE) + P(no\ DE)P(A|no\ DE)}$

$= \dfrac{.80(.32)}{.80(.32) + .20(.55)} = .699$

Section 6.6

17. Let H = high school diploma, NH = no high school diploma, and U = unemployed. The given information is

$$P(H) = .82, \quad P(NH) = .18, \quad P(U|H) = .04, \quad P(U|NH) = .10$$

We are to find $P(NH|U)$.

$$P(NH|U) = \frac{P(NH)P(U|NH)}{P(NH)P(U|NH) + P(H)P(U|H)}$$

$$= \frac{.18(.10)}{.18(.10) + .82(.04)} = .354$$

The probability the person has no high school diploma is .35.

19. $$P(hep|react) = \frac{(.70)(.95)}{(.70)(.95) + (.30)(.02)} = \frac{.665}{.671} = 0.991$$

21. Let 1.5+ represent those who spent at least 1.5 hours on homework and 1.5- those who spent less.

We are given $P(1.5+) = .43, \quad P(1.5-) = .57,$

$$P(A \text{ or } B \mid 1.5+) = .78, \quad P(A \text{ or } B|1.5-) = .21$$

Find $P(1.5+ \mid A \text{ or } B)$

$$P(1.5+ \mid A \text{ or } B) = \frac{P(1.5+) \ P(A \text{ or } B \mid 1.5+)}{P(1.5+)P(A \text{ or } B| \ 1.5+) + P(1.5-)P(A \text{ or } B \mid 1.5-)}$$

$$= \frac{.43(.78)}{.43(.78) + .57(.21)} = .737$$

The probability an A or B student studied at least 1.5 hours is about .737.

23. (a) $P(fraud) = (.11)(.20) + (.89)(.03) = 0.0487 = 4.87\%$

(b) $$P(exceed|fraud) = \frac{P(exceed \cap fraud)}{P(fraud)} = \frac{(.11)(.20)}{.0487} = .452$$

25. We are given

$$P(So) = .20, \quad P(Jr) = .35, \quad P(Sr) = .45$$
$$P(Ac|So) = .15, \quad P(Ac|Jr) = .08, \quad P(Ac|Sr) = .21$$

Find $P(Jr|Ac)$.

$$P(Jr|Ac) = \frac{P(Jr)P(Ac|Jr)}{P(So)P(Ac|So) + P(Jr)P(Ac|Jr) + P(Sr)P(Ac|Sr)}$$

$$= \frac{.35(.08)}{.20(.15) + .35(.08) + .45(.21)} = .184$$

About 18.4% of accounting majors were juniors.

27. We are given

$$P(R) = .54, \quad P(D) = .45, \quad P(I) = .01$$
$$P(Support|R) = .20, \quad P(Support|D) = .85, \quad P(Support|I) = 1$$

Find $P(R|Support)$

$$P(R|Support) = \frac{P(R)P(Support|R)}{P(R)P(Support|R) + P(D)P(Support|D) + P(I)P(Support(I))}$$

$$= \frac{.54(.20)}{.54(.20) + .45(.85) + .01(1)} = .216$$

The probability a person who supports the amendment is a Republican is .216.

29. $$P(Y|def) = \frac{(.35)(5/350)}{(.40)(1/40) + (.35)(5/350) + (.25)(2/250)} = \frac{.005}{.017} = .294$$

135

31.   We are given the following information
            p(perform satisfactorily | pass test) = .90
            p(not perform satisfactorily | pass test) = .10
            p(perform satisfactorily | not pass test) = .15
            p(not perform satisfactorily | not pass test) = .85
            p(passes test) = .65
    We wish to find
            p(pass test | perform satisfactorily)
    We draw a tree diagram that shows the information given.

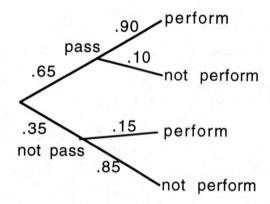

    Bayes' Rule states

$$p(pass \mid performs) = \frac{p(pass\ and\ performs)}{p(performs)}$$

    From the diagram there are two paths which terminate at performs
    satisfactorily so the probability of performing satisfactorily is
    the sum of those probabilities, that is,
            .65(.90) + .35(.15) = .6375
    p(pass and perform satisfactorily) = .65(.90) = .585

    Thus, p(pass | performs satisfactorily) = $\frac{.585}{.6375}$ = .918.

Section 6.7

1.   10(.0429)(.4225) = .181   3.   210(.0256)(.0466) = .251

5.   792(.1160)(.00064) = .059

7. $C(8, 3)(.25)^3(.75)^5 = 56(.0153)(.2373) = .208$

9. $C(5, 4)(0.1)^4(0.9)^1 = 0.00045$

11. $P(X = 3) = C(5, 3)(.25)^3(.75)^2 = 0.0879$

13.

X	P(X successes)
0	.240
1	.412
2	.265
3	.0756
4	.0081

15.

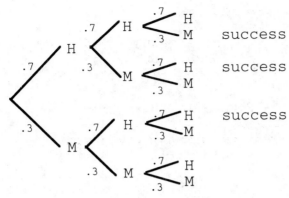

success

success

success

$P(\text{two hits in 3 attempts}) = 3(.7)(.7)(.3) = 0.44$

17. $n = 5$, $p = .8$, $x = 3$

$P(x = 3) = C(5, 3)(.8)^3(.2)^2 = 0.205$

19. $n = 3$, $P = 1/4$, $x = 2$

$P(x = 2) = C(3, 2)(1/4)^2(3/4)^1 = 0.141$

21. $n = 6$, $P = 1/6$ $P(x = 4) = C(6, 4)(1/6)^4(5/6)^2 = 0.0080$

23. (a) $n = 4$, $p = 1/6$, $P(x = 0) = C(4, 0)(1/6)^0(5/6)^4 = 0.482$

   (b) $n = 4$, $p = 1/6$, $P(x = 1) = C(4, 1)(1/6)^1(5/6)^3 = 0.386$

   (c) $n = 4$, $p = 1/6$, $P(x = 2) = C(4, 2)(1/6)^2(5/6)^2 = 0.116$

   (d) $n = 4$, $p = 1/6$, $P(x = 3) = C(4, 3)(1/6)^3(5/6)^1 = 0.0154$

   (e) $n = 4$, $p = 1/6$, $P(x = 4) = C(4, 4)(1/6)^4(5/6)^0 = 0.00077$

25. $n = 6$, $p = 1/2$, $P(x = 4) = C(6, 4)(1/2)^4(1/2)^2 = 0.234$

27. This is a repeated trials with $n = 6$, $x = 4$, $p = .8$, and $q = .2$
$P(x = 4) = C(6, 4)(.8)^4(.2)^2 = 15(.4096)(.04) = .246$
The probability 4 of the next 6 customers will tip 15% or more is .246.

29. $n = 5$ and $p = .5$, $q = .5$

X	P(X heads)
0	.0313
1	.156
2	.313
3	.313
4	.156
5	.0313

31. This is a repeated trials with $n = 5$, $x = 3$, $p = .4$, and $q = .6$
$P(x = 3) = C(5,3)(.4)^3(.6)^2 = 10(.064)(.36) = .230$
The probability Meka will receive more than 40% for three books is .230.

33. This is repeated trials with $n = 5$, $p = .6$, and $q = 4$.
We are to find $P(x = 3) + P(x = 4) + P(x = 5)$ which is
$C(5, 3)(.6)^3(.4)^2 + C(5, 4)(.6)^4(.4) + C(5, 5)(.6)^5$
$= 10(.216)(.16) + 5(.1296)(.4) + 1(.07776) = .683$
The probability at least three will raise their HDL level by 20% is about .68.

137

35. $n = 5$, $p = 1/6$
$P(x \geq 3) = P(x = 3) + P(x = 4) + P(x = 5) =$
$C(5, 3)(1/6)^3(5/6)^2 + C(5, 4)(1/6)^4(5/6)^1 + C(5, 5)(1/6)^5(5/6)^0$
$= 0.0355$

37. $n = 8$, $p = 1/2$,
$P(x \geq 5) = P(x = 5) + P(x = 6) + P(x = 7) + P(x = 8)$
$= C(8, 5)(1/2)^5(1/2)^3 + C(8, 6)(1/2)^6(1/2)^2$
$+ C(8, 7)(1/2)^7(1/2) + C(8, 8)(1/2)^8 = 0.363$

39.   (a) $n = 10$, $p = .5$, $P(x = 8) = C(10, 8)(.5)^8(.5)^2 = 0.0439$
       (b) $n = 10$, $p = .5$,
$P(\geq 8) = P(x = 8) + P(x = 9) + P(x = 10) =$
$C(10,8)(.5)^8(.5)^2 + C(10,9)(.5)^9(.5)^1 + C(10,10)(.5)^0 = 0.0547$
       (c) $n = 10$, $p = .5$,
$P(x \leq 2) = P(x = 0) + P(x = 1) + P(x = 2) =$
$C(10,0)(.5)^0(.5)^{10} + C(10,1)(.5)^1(.5)^9 + C(10,2)(.5)^2(.5)^8$
$= 0.0547$

41. $n = 6$, $p = 1/6$ $P(x = 4) = C(6, 4)(1/6)^4(5/6)^2 = 0.0080$

43. $n = 5$, $p = .36$
$P(x \geq 2) = 1 - P(x < 2) = 1 - [P(x = 0) + P(x = 1)]$
$= 1 - [C(5, 0)(.36)^0(.64)^5 + C(5, 1)(.36)^1(.64)^4] = 0.591$

45.   (a) $n = 4$, $p = .357$, $q = .643$
The probability of at least one hit is
$P(x = 1) + P(x = 2) + P(x = 3) + P(x = 4)$
$= C(4,1)(.357)^1(.643)^3 + C(4,2)(.357)^2(.643)^2$
$+ C(4,3)(.357)^3(.643) + C(4,4)(.357)^4$
$= 4(.357)(.2658) + 6(.1274)(.4134) + 4(.0455)(.643) + .01624$
$= .3796 + .3160 + .1170 + .01624 = .8288$
The probability of at least one hit is about .83.
       (b)   The probability of at least one hit in 56 consecutive games
is $(.83)^{56} = .000029$
This feat, as unlikely as it is, is considered by some to be
the greatest accomplishment in baseball.

47. $p = \dfrac{1}{2}$, $n = 10$, $X = 5$

$P(X = 5) = C(10, 5)(.5)^5(.5)^5 = .246$

49.   (a) $n = 4$, $p = .6$ (b) $n = 4$, $p = .4$

x	P(x successes)
0	.0256
1	.1536
2	.3456
3	.3456
4	.1296

x	P(x successes)
0	.1296
1	.3456
2	.3456
3	.1536
4	.0256

(c) n = 5, p = .5

x	P(x successes)
0	.03125
1	.15625
2	.31250
3	.31250
4	.15625
5	.03125

(d) n = 8, p = .3

x	P(x successes)
0	.0576
1	.1977
2	.2965
3	.2541
4	.1361
5	.0467
6	.0100
7	.0012
8	.00007

51. $C(220, 75)(.65)^{75}(.35)^{145} = 8.3 \times 10^{-21}$, practically zero.

53. (a) $P(x \geq 1) = 1 - P(x = 0) = 1 - (.8)^{30} = .9988$
     (b) $P(x \geq 5) = 1 - P(x = 0) - P(x = 1) - P(x = 2) - P(x = 3) -$
          $P(x = 4) =$
          $1 - (.8)^{30} - C(30, 1)(.2)(.8)^{29} - C(30, 2)(.2)^2(.8)^{28} -$
          $C(30, 3)(.2)^3(.8)^{27} - C(30, 4)(.2)^4(.8)^{26} = .745$

Section 6.8

1.   (a) Is a transition matrix.
     (b) Not a transition matrix because row 2 does not add to 1.
     (c) Not a transition matrix because it is not square.
     (d) A transition matrix.

3.   (a) 10%   (b) .60   (c) .90   (d) 40%

5.   (a) .4   (b) .8   (c) .3

7. $ST = [.675 \quad .325]$

9. $M_1 = M_0T = [.282 \quad .718]$
     $M_2 = M_1T = [.282 \quad .718]T = [.32616 \quad .67384]$

11. $M_1 = M_0T = [.175 \quad .375 \quad .45]$ $M_2 = M_1T = [.18 \quad .4075 \quad .4125]$

13. $ST = [2/3 \quad 1/3] \begin{bmatrix} .6 & .4 \\ .8 & .2 \end{bmatrix} = S$

15. $ST = [.464286 \quad .535714] \begin{bmatrix} .25 & .75 \\ .65 & .35 \end{bmatrix} = [13/28 \quad 15/28] = S$

17. $MT = [.25 \quad .33 \quad .42]$
     $(MT)T = [.259 \quad .323 \quad .418]$
     $((MT)T)T = [.2577 \quad .3233 \quad .419]$
     $MT^3 = [.2577 \quad .3233 \quad .419]$

19. Since $.2 + x + .4 = 1$, $x = .4$

21. Next year: $[.45 \quad .55] \begin{bmatrix} .75 & .25 \\ .15 & .85 \end{bmatrix} = [.42 \quad .58]$
                  so 42% will contribute next year
     In two years: $[.42 \quad .58] \begin{bmatrix} .75 & .25 \\ .15 & .85 \end{bmatrix} = [.402 \quad .598]$
                  so 40.2% will contribute in two years

23. $[.6 \quad .4] \begin{bmatrix} .5 & .5 \\ .2 & .8 \end{bmatrix} = [.38 \quad .62]$

$[.38 \quad .62] \begin{bmatrix} .5 & .5 \\ .2 & .8 \end{bmatrix} = [.314 \quad .686]$

$[.314 \quad .686] \begin{bmatrix} .5 & .5 \\ .2 & .8 \end{bmatrix} = [.2942 \quad .7058]$

$[.2942 \quad .7058] \begin{bmatrix} .5 & .5 \\ .2 & .8 \end{bmatrix} = [.28826 \quad .71174]$

$[.28826 \quad .71174] \begin{bmatrix} .5 & .5 \\ .2 & .8 \end{bmatrix} = [.286478 \quad .713522]$

$[.286478 \quad .713522] \begin{bmatrix} .5 & .5 \\ .2 & .8 \end{bmatrix} = [.2859434 \quad .7140566]$

It appears that $[.285 \quad .714]$ is the steady-state matrix.
The steady-state matrix is actually $[2/7 \quad 5/7]$.

25. Find $[x \quad y \quad z]$ such that
$$[x \quad y \quad z] \begin{bmatrix} .6 & .2 & .2 \\ .1 & .8 & .1 \\ .2 & .4 & .4 \end{bmatrix} = [x \quad y \quad z]$$

This with $x + y + z = 1$ gives

$$\begin{aligned} x + \quad y + \quad z &= 1 \\ -.4x + .1y + .2z &= 0 \\ .2x - .2y + .4z &= 0 \\ .2x + .1y - .6z &= 0 \end{aligned}$$

$$\begin{bmatrix} 1 & 1 & 1 & 1 \\ -.4 & .1 & .2 & 0 \\ .2 & -.2 & .4 & 0 \\ .2 & .1 & -.6 & 0 \end{bmatrix} \qquad \begin{bmatrix} 1 & 1 & 1 & 1 \\ 0 & .5 & .6 & .4 \\ 0 & -.4 & .2 & -.2 \\ 0 & -.1 & -.8 & -.2 \end{bmatrix}$$

$$\begin{bmatrix} 1 & 0 & -7 & -1 \\ 0 & 0 & -3.4 & -.6 \\ 0 & 0 & 3.4 & .6 \\ 0 & 1 & 8 & 2 \end{bmatrix} \qquad \begin{bmatrix} 1 & 0 & 0 & 4/17 \\ 0 & 1 & 0 & 10/17 \\ 0 & 0 & 1 & 3/17 \\ 0 & 0 & 0 & 0 \end{bmatrix}$$

so $[4/17 \quad 10/17 \quad 3/17]$ is the steady-state matrix.

27. $[x \quad y \quad z] \begin{bmatrix} .3 & .2 & .5 \\ .4 & .6 & 0 \\ .1 & .8 & .1 \end{bmatrix} = [x \quad y \quad z]$

and $x + y + z = 1$ gives the augmented matrix

$$\begin{bmatrix} 1 & 1 & 1 & 1 \\ -.7 & .4 & .1 & 0 \\ .2 & -.4 & .8 & 0 \\ .5 & 0 & -.9 & 0 \end{bmatrix} \qquad \begin{bmatrix} 1 & 1 & 1 & 1 \\ 0 & 1.1 & .8 & .7 \\ 0 & -.6 & .6 & -.2 \\ 0 & -.5 & -1.4 & -.5 \end{bmatrix}$$

Divide row 3 by $-.6$ and then interchange rows 2 and 3.

$$\begin{bmatrix} 1 & 1 & 1 & 1 \\ 0 & 1 & -1 & 1/3 \\ 0 & 1.1 & .8 & .7 \\ 0 & -.5 & -1.4 & -.5 \end{bmatrix} \qquad \begin{bmatrix} 1 & 0 & 2 & 2/3 \\ 0 & 1 & -1 & 1/3 \\ 0 & 0 & 1.9 & 1/3 \\ 0 & 0 & -1.9 & -1/3 \end{bmatrix}$$

$$\begin{bmatrix} 1 & 0 & 0 & 18/57 \\ 0 & 1 & 0 & 29/57 \\ 0 & 0 & 1 & 10/57 \\ 0 & 0 & 0 & 0 \end{bmatrix}$$

so $[18/57 \quad 29/57 \quad 10/57]$ is the steady-state matrix.

29. $[x \quad y \quad z] \begin{bmatrix} .6 & .2 & .2 \\ .1 & .8 & .1 \\ .2 & .3 & .5 \end{bmatrix} = [x \quad y \quad z]$ and $x + y + z = 1$ gives

$$\left[\begin{array}{ccc|c} 1 & 1 & 1 & 1 \\ -.4 & .1 & .2 & 0 \\ .2 & -.2 & .3 & 0 \\ .2 & .1 & -.5 & 0 \end{array}\right] \qquad \left[\begin{array}{ccc|c} 1 & 1 & 1 & 1 \\ 0 & .5 & .6 & .4 \\ 0 & -.4 & .1 & -.2 \\ 0 & -.1 & -.7 & -.2 \end{array}\right]$$

$$\left[\begin{array}{ccc|c} 1 & 0 & -1/5 & 1/5 \\ 0 & 1 & 6/5 & 4/5 \\ 0 & 0 & 29/50 & 3/25 \\ 0 & 0 & -29/50 & -3/25 \end{array}\right] \qquad \left[\begin{array}{ccc|c} 1 & 0 & 0 & 7/29 \\ 0 & 1 & 0 & 16/29 \\ 0 & 0 & 1 & 6/29 \\ 0 & 0 & 0 & 0 \end{array}\right]$$

The steady-state matrix = $[7/29 \quad 16/29 \quad 6/29]$ so at steady-state 24.14% are at I, 55.17% are at II, 20.69% are at III.

31. $[x \quad y] \begin{bmatrix} .9 & .1 \\ .1 & 0 \end{bmatrix} = [x \quad y]$ and $x + y = 1$ gives

$$\left[\begin{array}{cc|c} 1 & 1 & 1 \\ -.1 & .1 & 0 \\ .1 & -.1 & 0 \end{array}\right] \qquad \left[\begin{array}{cc|c} 1 & 1 & 1 \\ 0 & 1.1 & .1 \\ 0 & -1.1 & -.1 \end{array}\right] \qquad \left[\begin{array}{cc|c} 1 & 0 & 10/11 \\ 0 & 1 & 1/11 \\ 0 & 0 & 0 \end{array}\right]$$

so $[10/11 \quad 1/11]$ is the steady-state matrix.

33. $[x \quad y \quad z] \begin{bmatrix} 1/3 & 1/3 & 1/3 \\ 1/2 & 1/2 & 0 \\ 0 & 1/4 & 3/4 \end{bmatrix} = [x \quad y \quad z]$ and $x + y + z = 1$ gives

$$\left[\begin{array}{ccc|c} 1 & 1 & 1 & 1 \\ -2/3 & 1/2 & 0 & 0 \\ 1/3 & -1/2 & 1/4 & 0 \\ 1/3 & 0 & -1/4 & 0 \end{array}\right] \qquad \left[\begin{array}{ccc|c} 1 & 1 & 1 & 1 \\ 0 & 7/6 & 2/3 & 2/3 \\ 0 & -5/6 & -1/12 & -1/3 \\ 0 & -1/3 & -7/12 & -1/3 \end{array}\right]$$

$$\left[\begin{array}{ccc|c} 1 & 0 & 3/7 & 3/7 \\ 0 & 1 & 4/7 & 4/7 \\ 0 & 0 & 11/28 & 1/7 \\ 0 & 0 & -11/28 & -1/7 \end{array}\right] \qquad \left[\begin{array}{ccc|c} 1 & 0 & 0 & 3/11 \\ 0 & 1 & 0 & 4/11 \\ 0 & 0 & 1 & 4/11 \\ 0 & 0 & 0 & 0 \end{array}\right]$$

The steady-state matrix is $[3/11 \quad 4/11 \quad 4/11]$.

35.
$$\begin{array}{c} \\ \text{From} \end{array} \begin{array}{c} \\ 1 \\ 2 \\ 3 \\ 4 \\ 5 \end{array} \overset{\begin{array}{ccccc} & & \text{To} & & \\ 1 & 2 & 3 & 4 & 5 \end{array}}{\left[\begin{array}{ccccc} 0 & 1/2 & 0 & 0 & 1/2 \\ 1/2 & 0 & 1/2 & 0 & 0 \\ 0 & 1/2 & 0 & 1/2 & 0 \\ 0 & 0 & 1/2 & 0 & 1/2 \\ 1/2 & 0 & 0 & 1/2 & 0 \end{array}\right]}$$

37.
$$\left[\begin{array}{ccc|c} 1 & 1 & 1 & 1 \\ -1/2 & 1/4 & 0 & 0 \\ 1/2 & -1/2 & 1/2 & 0 \\ 0 & 1/4 & -1/2 & 0 \end{array}\right] \qquad \left[\begin{array}{ccc|c} 1 & 1 & 1 & 1 \\ 0 & 3/4 & 1/2 & 1/2 \\ 0 & -1 & 0 & -1/2 \\ 0 & 1/4 & -1/2 & 0 \end{array}\right]$$

$$\left[\begin{array}{ccc|c} 1 & 1 & 1 & 1 \\ 0 & -1 & 0 & -1/2 \\ 0 & 3/4 & 1/2 & 1/2 \\ 0 & 1/4 & -1/2 & 0 \end{array}\right] \qquad \left[\begin{array}{ccc|c} 1 & 0 & 1 & 1/2 \\ 0 & 1 & 0 & 1/2 \\ 0 & 0 & 1/2 & 1/8 \\ 0 & 0 & -1/2 & -1/8 \end{array}\right]$$

$$\begin{bmatrix} 1 & 0 & 0 & \bigm| & 1/4 \\ 0 & 1 & 0 & \bigm| & 1/2 \\ 0 & 0 & 1 & \bigm| & 1/4 \\ 0 & 0 & 0 & \bigm| & 0 \end{bmatrix}$$

When the process reaches a steady state 25% of the flowers will be red, 50% of the flowers will be pink, and 25% of the flowers will be white.

39.   (a)   .06   (b)   .05   (c)   .80   (d)   .04

Review Exercises,   Chapter 6

1.   No, because the probability assignments do not sum to 1.

3.   P(medium-sized soft drink) = 146/360 = .41 since a total of 360 soft drinks were sold.

5.   P(sports or band) = P(sports) + P(band) - P(both)
              = 1/3 + 2/5 - 1/6 = 17/30 = 0.57

7.   Four students can be selected in C(10, 4) = 210 ways. Mark, Melanie, and two others can be selected in C(2, 2)C(8, 2) = 28 ways.
     The probability Mark and Melanie are among the four selected is $\dfrac{28}{210} = \dfrac{2}{15} = .13$.

9.   P(king or spade) = P(king) + P(spade) - P(king of spades)
              = 4/52 + 13/52 - 1/52 = 4/13

11.   Let heads be considered a success. This is a Bernoulli experiment with n = 5, p = 1/2, and x = 5
              $P(x = 5) = C(5, 5)(1/2)^5(1/2)^0 = 1/32$

13.   (a)   P(red or white ball) = P(red) + P(white)
                          = 5/12 + 4/12 = 9/12 = 3/4
      (b)   P(2 black or 2 red) = P(2 black) + P(2 red)
                          = (3/12)(2/11) + (5/12)(4/11)
                          = (1/22) + (5/33) = 13/66 = 0.20

15.   (a) $P(\text{same kind}) = \dfrac{C(10,\ 2)}{C(40,\ 2)} + \dfrac{C(10,\ 2)}{C(40,\ 2)} + \dfrac{C(10,\ 2)}{C(40,\ 2)} + \dfrac{C(10,\ 2)}{C(40,\ 2)}$
                          = 4(45/780) = 0.23
      (b)   P(different kinds) = 1 - P(same kinds) = 1 - 0.23 = 0.77

17. $C(4,\ 2)(.4)^2(.6)^2 = (6)(.16)(.36) = .3456$

19.   P(Even) = 20/52   P(Ten) = 4/52
      P(Even) P(Ten) = (20/52)(4/52) = 5/169   P(Even ∩ Ten) = 4/52
      Since P(Even) P(Ten) ≠ P(Even ∩ Ten), Even and Ten are not independent events

21. (a) P(1, 2, 3, 4, in that order) = (1/6)(1/6)(1/6)(1/6) = 1/1296
     (b) P(1, 2, 3, 4, in any order) = Any one of the four on the
          first roll, any of the 3 remaining on second roll, etc.
          = (4/6)(3/6)(2/6)(1/6) = 24/1296 = 1/54
     (c) P(two even, then a 5, then a number less than 3)
          = (3/6)(3/6)(1/6)(2/6) = 18/1296 = 1/72

23. (a) P(passing math, history, and English, and failing chemistry)
          = P(passing math) × P(passing history)
              × P(passing English) × P(failing chemistry)
          = (.8)(.3)(.5)(.3) = 0.036
     (b) P(passing math and chemistry and failing history and English)
          = P(passing math) × P(passing chemistry)
              × P(failing history) × P(failing English)
          = (.8)(.7)(.7)(.5) = 0.196
     (c) P(passing all 4 subjects) = (.8)(.5)(.3)(.7) = 0.084

25. (a) P(male|abstainer)

$$= \frac{P(\text{male})P(\text{abstain}|\text{male})}{P(\text{male})P(\text{abstain}|\text{male}) + P(\text{female})P(\text{abstain}|\text{female})}$$

$$= \frac{(.45)(.20)}{(.45)(.20) + (.55)(.40)} = \frac{0.09}{0.31} = 0.29$$

     (b) P(female|heavy drinker)

$$= \frac{P(\text{female})P(\text{heavy drinker}|\text{female})}{P(\text{female})P(\text{heavy drink}|\text{female}) + P(\text{male})P(\text{heavy drink}|\text{male})}$$

$$= \frac{(.55)(.05)}{(.55)(.05) + (.45)(.20)} = \frac{0.0275}{0.1175} = 0.23$$

27. (a) 0 since repetitions are not allowed
     (b) 1/30   (c)   P(64 or 46) = 2/30
     (d)

First digit is	Number of 2-digit numbers
1	0
2	1
3	2
4	3
5	4
6	<u>5</u>
	15

          There are 15 possible numbers so p = 15/30
     (e) 7/30

29. Let J, S, and G be the event of being a junior, senior, and
     grad student, respectively. Let A be the event of receiving an A.
     P(J) = 10/50 = .2   P(A|J) = 2/10 = .2
     P(S) = 34/50 = .68   P(A|S) = 8/34 = .235
     P(G) = 6/50 = .12   P(A|G) = 3/6 = .5

$$P(J|A) = \frac{P(J)\ P(A|J)}{P(J)\ P(A|J) + P(S)\ P(A|S) + P(G)\ P(A|G)}$$

$$= \frac{(.2)(.2)}{(.2)(.2) + (.68)(.235) + (.12)(.5)} = 0.154$$

31. This is a Bernoulli experiment with n = 8, p = 1/3
     $P(x = 6) = C(8, 6)(1/3)^6 (2/3)^2 = (28)(.00137)(.444) = 0.017$

143

33. (a) $2 \times 1 \times 3 \times 2 \times 1 = 12$
    (b) A semifinalist must be seated first, then a finalist, etc.
        $3 \times 2 \times 2 \times 1 \times 1 = 12$

35. (a) $P(\text{First} < 3 \text{ and second} > 7) = (\frac{2}{10})(\frac{3}{10}) = .06$

    (b) $P(\text{First} < 4 \text{ and second} < 4) = (\frac{3}{10})(\frac{3}{10}) = .09$

37. (a) $P(\text{C or above}) = \frac{225}{420} = .536$ (b) $P(\text{Low}) = \frac{68}{420} = .162$

    (c) $P(\text{Below C} \mid \text{Middle}) = \frac{118}{242} = .488$

    (d) $P(\text{High} \mid \text{C or above}) = \frac{98}{225} = .436$

39. $P(\text{Seat belt not used}) = \frac{94}{192}$ $P(\text{Injuries}) = \frac{80}{192}$

    $P(\text{Seat belt not used and injuries}) = \frac{66}{192} = .344$

    $P(\text{Seat belt not used})P(\text{Injuries}) = \left(\frac{94}{192}\right)\left(\frac{80}{192}\right) = .204$

    Since $.344 \neq .204$ injuries and seat belt not used are dependent.

Chapter 7
Statistics

Section 7.1

1.
	Number	Frequency
(a)	-1	2
(b)	2	3
(c)	4	4
(d)	6	2
(e)	8	1

3. Number of students who used swimming pool per day

Classes	Frequency
85-99	7
100-114	8
115-129	13
130-144	7
145-159	5

5. (a) 67 students took the test.
    (b) 7 + 12 = 19 students scored below 70.
    (c) 14 + 8 = 22 students scored at least 80.
    (d) Cannot be determined.
    (e) 26 + 14 = 40 students scored between 69 and 90.
    (f) Cannot be determined.

7.
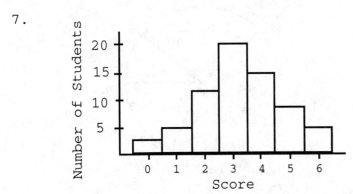

9. Since the interval 0-10 includes all scores, use intervals of
    length 2 and count the number in each interval.

Interval	Frequency
0-2.0	3
2.01-4.0	15
4.01-6.0	7
6.01-8.0	3
8.01-10.0	2

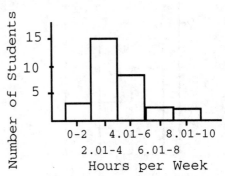

11.

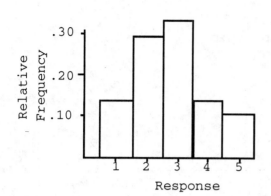

13.

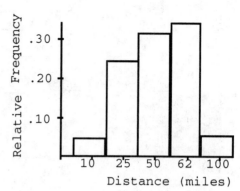

15.

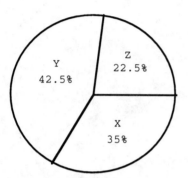

Brand of coffee preferred

17.

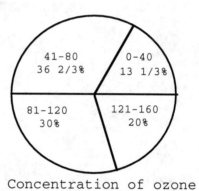

Concentration of ozone

19.

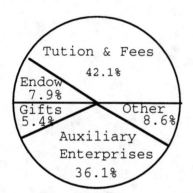

Income of Old Main University

21.

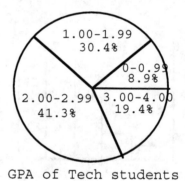

GPA of Tech students

25.

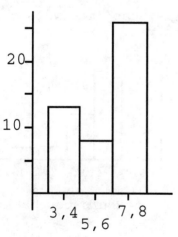

Section 7.2

27.

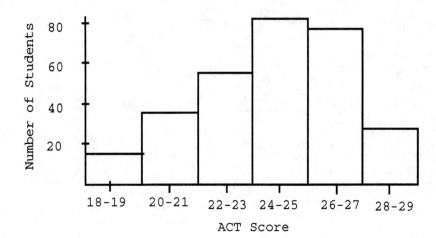

29.

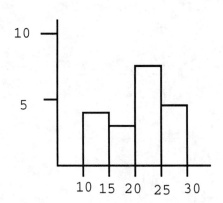

Section 7-2

1. $\mu = \dfrac{2 + 4 + 6 + 8 + 10}{5} = 6$

3. $\mu = \dfrac{2.1 + 3.7 + 5.9}{3} = 3.9$

5. $\mu = \dfrac{6 - 4 + 3 + 5 - 8 + 2}{6} = 0.67$

7. $\mu = \dfrac{5.9 + 2.1 + 6.6 + 4.7}{4} = 4.825$

9. $\mu = \dfrac{80 + 76 + 92 + 64 + 93 + 81 + 57 + 77}{8} = 77.5$

11. $\mu = \dfrac{90.5 + 89.2 + 78.4 + 91.0 + 84.2 + 73.5 + 88.7}{7} = 85.07$ mi/hr

13. 9   15. $(21 + 25)/2 = 23$

17. $(9 + 14)/2 = 11.5$ 19. Median is 72

21. 5   23. 1 and 5

25. No mode

27. $\dfrac{2 \times 96 + 3 \times 91 + 7 \times 85 + 13 \times 80 + 12 \times 75 + 10 \times 70 + 8 \times 60 + 5 \times 50}{60} = 73.83$

147

29. $$\frac{0 \times 430 + 1 \times 395 + 2 \times 145 + 3 \times 25 + 4 \times 5}{1000} = 0.78$$

31. $$\frac{2.5 \times 28 + 8.0 \times 47 + 15.5 \times 68 + 25.5 \times 7}{150} = 11.19$$

33. $$\frac{95 \times 8 + 84.5 \times 15 + 74.5 \times 22 + 64.5 \times 11 + 52 \times 5}{61} = 76$$

35. $$\frac{139.50 + 141.25 + 140.75 + 138.50 + 132.00}{5} = \$138.40$$

37. $$\frac{3.60 + 3.57 + 3.90 + 3.85 + 4.00 + 4.15 + 4.25 + 4.40}{8} = 3.965$$

Mean = $3.97
For median, arrange in order:

    3.57, 3.60, 3.85, 3.90, 4.00, 4.15, 4.25, 4.40

$$\frac{3.90 + 4.00}{2} = 3.95, \text{ median}$$

39.   (a)   5   (b)   2

      (c) $\dfrac{1 \times 5 + 2 \times 9 + 3 \times 3 + 4 \times 1}{18} = 2$ (d)   2

41.   Let x = the unknown grade

$$78 = \frac{72 + 88 + 81 + 67 + x}{5}$$

390 = 72 + 88 + 81 + 67 + x
x = 82

43. $$\frac{21 \times 18000 + 9 \times 22000 + 14 \times 25000 + 26 \times 28000 + 13 \times 30000 + 8 \times 35000}{91}$$

      = \$25,538.46

45. $$\frac{8(27450) + 10(31400)}{18} = \frac{533600}{18} = \$29,644.44$$

47.   Put the scores in order: 65, 77, 82, 93
Since there are an odd number of test scores, the median, 82, must be the third number in the ordered group of test scores. So the fifth test score must be 82 or larger.

49. $$\frac{3(1.27) + 2(1.34)}{5} = \frac{6.49}{5} = \$1.30$$

51. $$\mu = \frac{4(1) + 28(2) + 95(3) + 24(4) + 4(5)}{155} = 2.974$$

55.   (a)   Mean per capita income = $13,995.
              Mean number below poverty = 32.2 million

57.   69.35   59.   22.3

1. $\mu = \dfrac{19 + 10 + 15 + 20}{4} = 16$

    $\text{var} = \dfrac{(19 - 16)^2 + (10 - 16)^2 + (15 - 16)^2 + (20 - 16)^2}{4} = 15.5,$

    $\sigma = \sqrt{15.5} = 3.94$

3. $\mu = \dfrac{4 + 8 + 9 + 10 + 14}{5} = 9$

    $\text{var} = \dfrac{(4 - 9)^2 + (8 - 9)^2 + (9 - 9)^2 + (10 - 9)^2 + (14 - 9)^2}{5} = 10.4$

    $\sigma = \sqrt{10.4} = 3.22$

5. $\mu = \dfrac{17 + 39 + 54 + 22 + 16 + 46 + 25 + 19 + 62 + 50}{10} = 35$

    $(17 - 35)^2 + (39 - 35)^2 + (54 - 35)^2 + (22 - 35)^2 + (16 - 35)^2$
    $+ (46 - 35)^2 + (25 - 35)^2 + (19 - 35)^2 + (62 - 35)^2 + (50 - 35)^2$
    $= 2662$

    $\text{var} = \dfrac{2662}{10} = 266.2, \quad \sigma = \sqrt{266.2} = 16.32$

7. $\mu = \dfrac{-8 - 4 - 3 + 0 + 1 + 2}{6} = -2$

    $\text{var} =$
    $\dfrac{(-8 + 2)^2 + (-4 + 2)^2 + (-3 + 2)^2 + (0 + 2)^2 + (1 + 2)^2 + (2 + 2)^2}{6}$

    $= 11.67, \quad \sigma = \sqrt{11.67} = 3.42$

9. $\mu = \dfrac{-3 + 0 + 1 + 4 + 5 + 8 + 10 + 11}{8} = 36/8 = 4.5$

    $\text{var} =$
    $\dfrac{(-3-4.5)^2 + (0-4.5)^2 + (1-4.5)^2 + (4-4.5)^2 + (5-4.5)^2 + (8-4.5)^2 + (10-4.5)^2 + (11-4.5)^2}{8}$

    $= 174/8 = 21.75, \quad \sigma = 4.66$

11. $\mu = \dfrac{2 \times 1 + 5 \times 2 + 2 \times 3 + 3 \times 4 + 8 \times 5}{20} = 70/20 = 3.5$

    $\text{var} = \dfrac{2(1-3.5)^2 + 5(2-3.5)^2 + 2(3-3.5)^2 + 3(4-3.5)^2 + 8(5-3.5)^2}{20}$

    $= 43/20 = 2.15, \quad \sigma = 1.47$

13. $\mu = \dfrac{10(4.65) + 12(4.90) + 8(5.00) + 5(5.24)}{35} = 171.5/35 = 4.90$

    $\text{var} = \dfrac{10(-.25)^2 + 12(0)^2 + 8(.10)^2 + 5(.34)^2}{35}$

    $= 1.283/35 = 0.0367, \quad \sigma = 0.191$

15. $\mu = \dfrac{5(5) + 12(15.5) + 8(25.5)}{25} = 415/25 = 16.6$

    $\text{var} = \dfrac{5(-11.6)^2 + 12(-1.1)^2 + 8(8.9)^2}{25} = 1321/25 = 52.84$

    $\sigma = 7.27$

17. $\mu = \dfrac{14(12.5) + 18(18) + 18(23)}{50} = 913/50 = 18.26$

    $\sigma^2 = \dfrac{14(-5.76)^2 + 18(-.26)^2 + 18(4.74)^2}{50} = 17.4024$

    $\sigma = \sqrt{17.4024} = 4.17$

19. $\mu = 217/7 = 31$, $\sigma = \sqrt{68/7} = \sqrt{9.714} = 3.12$

21. (a) $z = (180 - 160)/16 = 20/16 = 1.25$
    (b) $z = -10/16 = -0.625$ (c) $z = 0/16 = 0$
    (d) $1 = (x - 160)/16$ so $x = 16 + 160 = 176$
    (e) $-0.875 = (x - 160)/16$ so $x = 16(-0.875) + 160 = 146$

23. (a) <u>Lathe I</u>
    variance =
    $$\dfrac{(.501-.5)^2 + (.503-.5)^2 + (.495-.5)^2 + (.504-.5)^2 + (.497-.5)^2}{5}$$
    $= 0.000012$
    $\sigma = \sqrt{0.000012} = 0.00346$, for Lathe I.
    <u>Lathe II</u>
    variance =
    $$\dfrac{(.502-.5)^2 + (.497-.5)^2 + (.498-.5)^2 + (.501-.5)^2 + (.502-.5)^2}{5}$$
    $= 0.000044$
    $\sigma = \sqrt{0.000044} = 0.00210$, for Lathe II.
    (b) Lathe II is more consistent because σ is smaller.

25. Let x be the number of 2's in the set.
    Let y be the number of 3's in the set.
    Let 8 - x - y be the number of 4's in the set.
    (a) $\dfrac{2x + 3y + 4(8 - x - y)}{8} = 3$

    $.5 = \sqrt{\dfrac{x(2-3)^2 + y(3-3)^2 + (8-x-y)(4-3)^2}{8}}$

    These can be rewritten as $2x + y = 8$ and
    $.25 = \dfrac{x + (8 - x - y)}{8}$ or $2 = 8 - y$
    so $y = 6$, the number of 3's
    $2x + y = 8$
    $2x + 6 = 8$
    $2x = 2$
    $x = 1$, the number of 2's
    $8 - x - y = 8 - 1 - 6 = 1$, the number of 4's
    The set of numbers is 2, 3, 3, 3, 3, 3, 3, 4
    (b) The first equation is the same, $2x + y = 8$ and the second is
    $1 = \dfrac{x + (8 - x - y)}{8}$ or $8 = 8 - y$
    so $y = 0$, the number of 3's
    $2x + 0 = 8$
    $x = 4$, the number of 2's
    $8 - x - y = 8 - 4 = 4$, the number of 4's
    The set of numbers is 2, 2, 2, 2, 4, 4, 4, 4

(c)   No, because the largest possible value of the sum of the squared deviations is 8, for which $\sigma = 1$.

27.   30% scored above her so $110(.30) = 33$ scored above.
She ranked 34th out of 110.

29.   2% scored higher so $.02(1545) = 30.9$ or 31 scored higher

31.   On the first test his z-score was $z_1 = (86 - 72)/8 = 14/8 = 1.75$, on the second test his z-score was $z_2 = (82 - 62)/12 = 20/12 = 1.67$
The 86 was the better score because he scored higher above the mean.

33. $z_R = (19 - 18)/1 = 1/1 = 1$
$z_C = (64 - 59)/3 = 5/3 = 1.7$
The runner's time was one z-score higher than the mean so the runner was slower than average. The cyclist's time was 1.67 z-scores above the mean so the cyclist was even slower than the average. The runner had the better performance.

37. $\mu = 19.26$ $\sigma = 3.79$ 39. $\mu = 3.5$ $\sigma = 1.47$

41. $\mu = 4.9$ $\sigma = .19$ 43. $\mu = 18.26$ $\sigma = 4.17$

45. $\mu = 3869.6$ $\sigma = 1181.9$

47. $\mu = 7.49$ million $\sigma = 1.95$ million

Section 7.4

1.
HHH	$X = 3$
HHT	$X = 2$
HTH	$X = 2$
THH	$X = 2$
TTH	$X = 1$
THT	$X = 1$
HTT	$X = 1$
TTT	$X = 0$

3.
Ann, Betty	$X = 2$
Ann, Jason	$X = 1$
Ann, Tom	$X = 1$
Betty, Jason,	$X = 1$
Betty, Tom	$X = 1$
Jason, Tom	$X = 0$

5. $X = 0, 1, 2,$ or 3

7.   (a) $X = 0, 1, 2, 3,$ or 4   (b) $X = 0, 1, 2,$ or 3

9.   (a)   Discrete   (b)   Continuous
(c)   Continuous   (d)   Discrete

11.   (a)   Continuous   (b)   Discrete
(c)   i) Discrete   ii) Continuous   iii) Discrete

13.   Yes, since sum is 1.

15.

X	P(X)
0	1/8
1	3/8
2	3/8
3	1/8

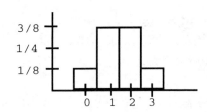

17.

X	P(X)
3	1/5
4	1/5
5	2/5
6	1/5

19.

X	P(X)
0	8/75
1	49/75
2	13/75
3	4/75
4	1/75

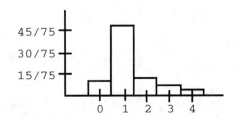

21.

	X	P(X)		
(Red and black)	0	$\dfrac{4(6)}{C(10,\ 2)}$	$= \dfrac{24}{45}$	$= .533$
(2 black)	5	$\dfrac{C(6,\ 2)}{C(10,\ 2)}$	$= \dfrac{15}{45}$	$= .333$
(2 red)	10	$\dfrac{C(4,\ 2)}{C(10,\ 2)}$	$= \dfrac{6}{45}$	$= .133$

23. (a) $X \in \{1, 2, 3, \dots\}$ (b) Discrete

25. $X = 0$: $C(5,\ 0) \cdot C(6,\ 2) = (1)(15) = 15$
     $X = 1$: $C(5,\ 1) \cdot C(6,\ 1) = (5)(6) = 30$
     $X = 2$: $C(5,\ 2) \cdot C(6,\ 0) = (10)(1) = 10$

27.

X	P(X)		
0	$\dfrac{C(4,\ 0)\ C(4,\ 2)}{C(8,\ 2)}$	$= \dfrac{6}{28}$	$= \dfrac{3}{14}$
1	$\dfrac{C(4,\ 1) \cdot C(4,\ 1)}{C(8,\ 2)}$	$= \dfrac{4 \cdot 4}{28}$	$= \dfrac{4}{7}$
2	$\dfrac{C(4,\ 2) \cdot C(4,\ 0)}{C(8,\ 2)}$	$= \dfrac{6}{28}$	$= \dfrac{3}{14}$

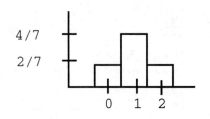

29.

X	P(X)		
0	$\dfrac{C(7,\ 2)}{C(10,\ 2)}$	$= \dfrac{21}{45}$	$= \dfrac{7}{15}$
1	$\dfrac{C(7,\ 1)\ C(3,\ 1)}{C(10,\ 2)}$	$= \dfrac{21}{45}$	$= \dfrac{7}{15}$
2	$\dfrac{C(3,\ 2)}{C(10,\ 2)}$	$= \dfrac{3}{45}$	$= \dfrac{1}{15}$

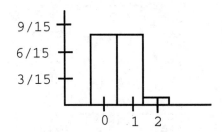

31.   (a)

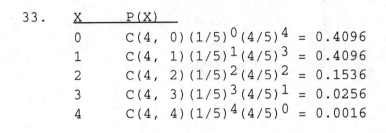

X	P(X)	
0	$\dfrac{C(4,\ 2)}{C(6,\ 2)}$	$= \dfrac{2}{5}$
1	$\dfrac{C(4,\ 1)C(2,\ 1)}{C(6,\ 2)}$	$= \dfrac{8}{15}$
2	$\dfrac{C(2,\ 2)}{C(6,\ 2)}$	$= \dfrac{1}{15}$

   (b)

X	P(X)	
0	$\dfrac{C(5,\ 2)}{C(6,\ 2)}$	$= \dfrac{2}{3}$
1	$\dfrac{C(5,\ 1)\ C(1,\ 1)}{C(6,\ 2)}$	$= \dfrac{1}{3}$

33.

X	P(X)	
0	$C(4,\ 0)(1/5)^0(4/5)^4$	$= 0.4096$
1	$C(4,\ 1)(1/5)^1(4/5)^3$	$= 0.4096$
2	$C(4,\ 2)(1/5)^2(4/5)^2$	$= 0.1536$
3	$C(4,\ 3)(1/5)^3(4/5)^1$	$= 0.0256$
4	$C(4,\ 4)(1/5)^4(4/5)^0$	$= 0.0016$

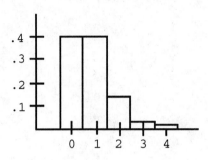

35.

X	P(X)	
0	$(7/12)^3$	$= 0.1985$
1	$C(3,\ 1)(5/12)^1(7/12)^2$	$= 0.4253$
2	$C(3,\ 2)(5/12)^2(7/12)^1$	$= 0.3038$
3	$(5/12)^3$	$= 0.0723$

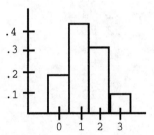

37.   X = 0, 1 sequence (Gold, Gold);
       X = 1, 2 sequences (Gold, Green, Gold; or Green, Gold, Gold)
       X = 2, 3 sequences (Green, Green; or Green, Gold, Green; or Gold,
                   Green, Green)

39.   (a)   Weight   (b)   Shirt size, suit size
       (c)   Shoe size   (d)   Number of cavities

41. $X \in \{1, 2, 3, 4, \ldots\}$

43.

X	P(X)
2	3/7
3	(4/7)(3/6) = 2/7
4	(4/7)(3/6)(3/5) = 6/35
5	(4/7)(3/6)(2/5)(3/4) = 3/35
6	(4/7)(3/6)(2/5)(1/4)(3/3) = 1/35

45.

X	P(X)	
1	1/2	(H)
2	(1/2)(1/2) = 1/4	(TH)
3	(1/2)(1/2)(1/2) + (1/2)(1/2)(1/2) = 1/4	(TTH or TTT)

47.

X	P(X)
0	$(.4)^2 = 0.16$
1	$2(.6)(.4)^2 = 0.192$ (BAB or ABB)
2	$(.6)^2 + 2(.6)^2(.4) = 0.648$ (AA or ABA or BAA)

49.

X(# cards drawn)	P(X)
1	0
2	5/11
3	3/11
4	4/33
5	5/66
6	5/154
7	5/231
8	2/231
9	1/154
10	1/462
11	1/462
12	1/462

(Cards alternate)

9	.0001

51.

X	P(X)
0	.0207
1	.1004
2	.2162
3	.2716
4	.2194
5	.1181
6	.0424
7	.0098
8	.0013

53.

X	P(X)
0	.0039
1	.0313
2	.1094
3	.2188
4	.2734
5	.2188
6	.1094
7	.0313
8	.0039

Section 7.5

1. $E(X) = .3(3) + .2(8) + .1(15) + .4(22) = 12.8$

3. $E(X) = .2(150) + .2(235) + .2(350) + .2(410) + .2(480) = 325$

5. $E(X) = (8)(1/2) + (2)(1/2) = \$5$

7. (a) $E(X) = (1)(1/2) + 0(1/2) = \$0.50$ so a charge of \$.50 will break even.
   (b) $\$1 + \$.50 = \$1.50$

9. (a) $(.15)(20) + (.65)(130) + (.20)(700) = \$227.50$
   (b) $(125)(\$227.50) = \$28,437.50$

11. $\mu = .4(100) + .5(140) + .1(210) = 131$
    $\sigma^2 = .4(-31)^2 + .5(9)^2 + .1(79)^2 = 1049 \qquad \sigma = 32.39$

13. $\mu = .4(10) + .2(30) + .3(50) + .1(90) = 34$
    $\sigma^2 = .4(-24)^2 + .2(-4)^2 + .3(16)^2 + .1(56)^2 = 624 \qquad \sigma = 24.98$

15. $X = 0, 1,$ or 2 for the number of defective toys that can be selected.
    $E(X) = 0 + (1) \cdot \dfrac{4 \cdot 11}{C(15, 2)} + (2) \cdot \dfrac{C(4, 2)}{C(15, 2)} = \dfrac{44}{105} + \dfrac{12}{105} = \dfrac{56}{105} = .533$

17. $E(X_A) = (.60)(50,000) + (.30)(0) + (.10)(-70,000) = \$23,000$
    $E(X_B) = (.55)(100,000) + (.45)(-60,000) = \$28,000$
    B appears to be more profitable.

19. $E(X) = .75(0) + .10(1) + .06(2) + .04(3) + .04(4) + .01(5) = 0.55$

21. (a) $E(X) = 1(3/10) + 2(2/10) + 3(2/10) + 4(1/10) + 5(2/10) = 2.7$
    (b) $\mu = 2.7$, $\sigma^2 = 2.21$, $\sigma = 1.49$

23. (a) $E(X) = 0 + (2)C(3, 1)(.4)^1(.6)^2 + (4)C(3, 2)(.4)^2(.6)^1$
    $\qquad + (6)C(3, 3)(.4)^3(.6)^0 = \$2.40$
    (b) $E(X) = 0 + (2)C(4, 1)(.4)^1(.6)^3 + (4)C(4, 2)(.4)^2(.6)^2$
    $\qquad + (6)C(4, 3)(.4)^3(.6)^1 + (8)C(4, 4)(.4)^4(.6)^0 = \$3.20$

25. (a) $E(X) = .05(-23) + .95(18) = \$15.95$
    (b) $(150,000)(15.95) = \$2,392,500$

27. Let x = amount received by a player.
    $E(X) = .3x + .7(-5) = .3x - 3.5$ which is -$.20 since the casino
    plans to average $.20 per game.
    $\qquad$ For $.3x - 3.5 = -.20 \qquad x = 11$
    The payoff for winning is $11.

29. (a) The probability a player wins $1.00 is $\frac{18}{38}$ and the

    probability of losing $1.00 is $\frac{20}{38}$ so the expected value is

    $E = \frac{18}{38}(1) + \frac{20}{38}(-1) = -.0526$

    The expected value is about negative 5 cents per play, so a
    player can expect an average loss of about 5 cents per play.
    (b) Since the expected value is negative to the player, the game
    favors the house.
    (c) For the game to be fair a payoff of x dollars would give
    $\qquad \frac{18}{38}x + \frac{20}{38}(-1) = 0$
    $\qquad 18x = 20$
    $\qquad\quad x = 1.1111$
    A payoff of $1.11 would make the game fair.
    (d) For a bet on green the probability of winning x dollars is
    $\frac{2}{38}$ and the probability of losing one dollar is $\frac{36}{38}$. To be a

    fair game $\frac{2}{38}x + \frac{36}{38}(-1) = 0$
    $\qquad$ so $2x = 36$
    $\qquad\quad x = 18$
    A payoff of $18 for winning a green bet makes the game fair.

31. $\begin{bmatrix} 4 & 8 & 10 & 6 & 3 \end{bmatrix} \begin{bmatrix} .15 \\ .10 \\ .25 \\ .30 \\ .20 \end{bmatrix} = \begin{bmatrix} 6.3 \end{bmatrix}$ The expected value is 6.3.

33. $\begin{bmatrix} 140 & 150 & -75 & -50 & 200 & -250 \end{bmatrix} \begin{bmatrix} .12 \\ .08 \\ .24 \\ .16 \\ .22 \\ .18 \end{bmatrix} = \begin{bmatrix} 1.8 \end{bmatrix}$

The expected value is 1.8.

Section 7.6

1. $z = (3.1 - 4.0)/0.3 = -3$

3. $z = (10.1 - 10.0)/2.0 = 0.05$

5. $z = (2.65 - 0)/1.0 = 2.65$

7. $A = 0.1915$ 9. $A = 0.0987$

11. $A = 0.3643$ 13. $A = 0.2734$

15. $z = 0.46$ so $A = 0.1772 = 17.72\%$

17. $z = 0.38$ so $A = 0.1480 = 14.8\%$

19. $z = -1.24$ so $A = 0.3925 = 39.25\%$

21. $z = -2.9$ so $A = 0.4981 = 49.81\%$

23. For $z = 1.25$, $A = 0.3944$ so the total area is $A = 2(0.3944) = 0.7888$

25. For each z, $A = 0.4861$ so the total area is $A = 2(0.4861) = 0.9722$

27. $z = 0.65$ $A = 2(0.2422) = 0.4844 = 48.44\%$

29. $z = 0.38$ $A = 2(0.1480) = 0.2960 = 29.6\%$

31. $A = A_1 + A_2 = 0.2258 + 0.3997 = 0.6255$

33. $A = A_1 + A_2 = 0.2881 + 0.4974 = 0.7855$

35. $A = 0.5 - 0.4032 = 0.0968$

37. $A = 0.5 - 0.4918 = 0.0082$

39. $1 - 2(0.3051) = 0.3898 = 38.98\%$

41. $1 - 2(0.4332) = 0.1336 = 13.36\%$

43. $z = (98 - 85)/5 = 2.60$, $A = 0.4953$

45. $z = (80 - 85)/5 = -1.00$, $A = 0.3413$

47. $z_1 = (220 - 226)/12 = -0.5$, $z_2 = (235 - 226)/12 = 0.75$ so $A_1 + A_2 = 0.1915 + 0.2734 = 0.4649$

49. $z_1 = (211 - 226)/12 = -1.25$, $z_2 = (241 - 226)/12 = 1.25$ so $2(A) = 2(0.3944) = 0.7888$

51. $z_1 = (144 - 140)/8 = 0.5$, $z_2 = (152 - 140)/8 = 1.5$ so $A_2 - A_1 = 0.4332 - 0.1915 = 0.2417$

53. $z_1 = (146 - 140)/8 = 0.75$, $z_2 = (156 - 140)/8 = 2$
      so $A_2 - A_1 = 0.4773 - 0.2734 = 0.2039$

55. $z_1 = (80 - 75)/5 = 1$, $z_2 = (85 - 75)/5 = 2$
      so $A_2 - A_1 = 0.4773 - 0.3413 = 0.1360$

57. $z = (76 - 75)/5 = 0.2$ so $0.5 - 0.0793 = 0.4207$

59. $z_1 = (70 - 75)/5 = -1$, $z_2 = (80 - 75)/5 = 1$
      so $1 - 2(0.3413) = 0.3174$

61. $z_1 = (155 - 168)/10 = -1.3$, $z_2 = (169 - 168)/10 = 0.1$
      so $A_1 + A_2 = 0.4032 + 0.0398 = 0.4430 = 44.3\%$

63. $z = (172 - 168)/10 = 0.4$ so $0.5 + 0.1554 = 0.6554 = 65.54\%$

65. $z = (173 - 168)/10 = 0.5$ so $A = 0.5 - 0.1915 = 0.3085 = 30.85\%$

67. $z = (184 - 168)/10 = 1.6$ so $A = 0.5 + 0.4452 = 0.9452 = 94.52\%$

69.   Since 8% of the scores are to the right of z, A = .5 - .08 = .42 is the area between the mean and z. The z that corresponds to A = .42 is z = 1.41

71.   Since 86% of the scores lie to the left of z, z is above the mean and 36% of the scores lie between the mean and z. The value of z that corresponds to A = .36 is z = 1.08

73.   Since 91% of the scores lie between z and -z, one-half of the scores, 45.5%, lie between the mean and z. The value of z that corresponds to A = .455 is z = 1.70.

75.   (a) $z = (3.75 - 3.15)/.75 = 0.8$
            so $0.5 - A = 0.5 - 0.2881 = 0.2119 = 21.19\%$
      (b)   0.2119

77.   (a) $z_1 = (120 - 110)/12 = 0.83$, $z_2 = (125 - 110)/12 = 1.25$
            so $A_2 - A_1 = 0.3944 - 0.2967 = 0.0977$
      (b) $z = (100 - 110)/12 = -0.83$
            so $0.5 - A = 0.5 - 0.2967 = 0.2033$
      (c) $z_1 = (105 - 110)/12 = -0.42$,
            $z_2 = (115 - 110)/12 = 0.42$ so $2(A) = 2(0.1628) = 0.3256$

79.   (a) $z_1 = (120 - 126)/6 = -1$, $z_2 = (132 - 126)/6 = 1$
            so $2(0.3413)(700) = 477.82$ which we round to 478.
      (b) $z = (114 - 126)/6 = -2$
            so $(.5 - 0.4773)(700) = 15.89$ which we round to 16
      (c) $z = (134 - 126)/6 = 1.33$ so $(.5 - 0.4082)(700) = 64.26$
                  rounded to 64

81.   (a)   The area between the mean and the A cutoff is .5 - .12 = .38 so z = 1.18. Let x = the cutoff score. Then $1.18 = \dfrac{x - 66}{17}$ and x = 85.975 rounded to 86.

   (b)   The area between the mean and the A cutoff is .5 - .06 = .44 so z = 1.56. Let x = the cutoff score. Then $1.56 = \dfrac{x - 66}{17}$ gives x = 92.52 which we round to 92.

83.   (a) $z_1 = -0.75$, $z_2 = 0.75$ so A = 2(0.2734) = 0.5468

      (b) $z_1 = 12/40 = 0.3$, $z_2 = 30/40 = 0.75$
            so A = 0.2734 - 0.1179 = 0.1555

      (c)   .50 score less than 300
            $z_1 = 26/40 = 0.65$ so A = 0.5 + 0.2422 = 0.7422.
            0.7422 of the students score less than 326

      (d)   Let x be the cutoff score. Since 10% of the area under the normal curve is above x, then A = .5 - .1 = 0.4 is the area between the mean and x. For A = .4, z = 1.28 so $1.28 = \dfrac{x - 300}{40}$ which gives x = 351.2. A student must score 352 or higher.

      (e) $z_1 = 12/40 = 0.3$, $z_2 = 24/40 = 0.6$ so the area between the two scores is 0.2258 - 0.1179 = 0.1079 which also is the probability.

85.   For z = 0.8, A = 0.2881 so the probability a student scores below 128 is 0.5 + 0.2881 = 0.7881 and the probability both score less is $(0.7881)^2 = 0.6211$

87.   For z = (1140 - 1050)/50 = 1.8, A = 0.4641 which gives 0.0359 as the area above 1140. The probability the device lasts at least 1140 hours is 0.0359.

89.   (a)   .4332 - .1915 = .2417   (b)   .4192 - .2258 = .1934
      (c)   .4032 - .2580 = .1452   (d)   .3849 - .2881 = .0968
      (e)   .3643 - .3159 = .0484   (f)   .3531 - .3289 = .0242
      (g)   .3438 - .3389 = .0049

91.   No, the mean is the midpoint of the data in a normal distribution.

93.   Yes, the mean and median are approximately equal and near the midpoint.

95.   As a whole, the curve does not have the symmetry needed for a normal distribution. The portion obtained by deleting the 7 rightmost bars is somewhat normal in shape so data in that region might be considered approximately normal.

97.   (a)   We want to find the area under the normal curve between 450 and 750.

For School A
$$z_1 = \frac{450 - 610}{150} = -1.07, \quad A_1 = .3577$$
$$z_2 = \frac{750 - 610}{150} = .93, \quad A = .3238$$

            The total area is .6815 or 68.15% of the total area.
            Number of students from school A = .6815(500) = 340.75 or 341 students.

For School B
$$z_1 = \frac{450 - 590}{150} = -.93, \quad A_1 = .3238$$

$$z_2 = \frac{750 - 590}{150} = 1.07, \; A_2 = .3577$$

The total area is .6815 and the number of students is .6815(500) = 340.75 or 341 students.

The same number of students from each school scored in the middle range, 450-750 so this does not distinguish the two schools.

(b) We need to find the area under the normal curve above 840.

For School A

$$z = \frac{840 - 610}{150} = 1.53, \; A = .5000 - .4370 = .0630$$

The number of students from school A is 0630(500) = 31.5. We round this to 32 students.

For School B

$$z = \frac{840 - 590}{150} = 1.67, \; A = .5000 - .4525 = .0475$$

The number of students from school B is .0475(500) = 23.75. We round this to 24 students.

These numbers indicate a lower performance by school B because they have 3/4 the number of students in the top 5% as does school A.

(c) We need to find the area below 360.

For school A

$$z = \frac{360 - 610}{150} = -1.67, \; A = .5000 - .4525 = .0475$$

The number of students is .0475(500) = 23.75 which we round to 24.

For school B

$$z = \frac{360 - 590}{150} = -1.53, \; A = .5000 - .4370 = .0630$$

The number of students is .0630(500) = 31.5 which we round to 32.

The number of students from each school is the reverse from part (b). Again, this indicates more students are less prepared from school B than from A.

Section 7.7

1.

X	P(X)	
0	$C(5,0)(.3)^0(.7)^5$	= 0.1681
1	$C(5,1)(.3)^1(.7)^4$	= 0.3602
2	$C(5,2)(.3)^2(.7)^3$	= 0.3087
3	$C(5,3)(.3)^3(.7)^2$	= 0.1323
4	$C(5,4)(.3)^4(.7)^1$	= 0.02835
5	$C(5,5)(.3)^5(.7)^0$	= 0.00243

3.

X	P(X)	
0	$C(5,0)(.4)^0(.6)^5$	= 0.0778
1	$C(5,1)(.4)^1(.6)^4$	= 0.2592
2	$C(5,2)(.4)^2(.6)^3$	= 0.3456
3	$C(5,3)(.4)^3(.6)^2$	= 0.2304
4	$C(5,4)(.4)^4(.6)^1$	= 0.0768
5	$C(5,5)(.4)^5(.6)^0$	= 0.0102

5.

X	P(X)	
0	$C(4,0)(1/6)^0(5/6)^4$	= 0.4823
1	$C(4,1)(1/6)^1(5/6)^3$	= 0.3858
2	$C(4,2)(1/6)^2(5/6)^2$	= 0.1157
3	$C(4,3)(1/6)^3(5/6)^1$	= 0.0154
4	$C(4,4)(1/6)^4(5/6)^0$	= 0.0008

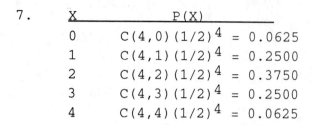

7.

X	P(X)	
0	$C(4,0)(1/2)^4$	= 0.0625
1	$C(4,1)(1/2)^4$	= 0.2500
2	$C(4,2)(1/2)^4$	= 0.3750
3	$C(4,3)(1/2)^4$	= 0.2500
4	$C(4,4)(1/2)^4$	= 0.0625

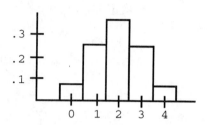

9.

X	P(X)	
0	$C(4,0)(.6)^0(.4)^4$	= 0.0256
1	$C(4,1)(.6)^1(.4)^3$	= 0.1536
2	$C(4,2)(.6)^2(.4)^2$	= 0.3456
3	$C(4,3)(.6)^3(.4)^1$	= 0.3456
4	$C(4,4)(.6)^4(.4)^0$	= 0.1296

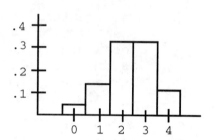

11.　　$P(x = 40) = C(90, 40)(0.4)^{40}(0.6)^{50}$

13.　　$P(x = 35) = C(75, 35)(0.25)^{35}(0.75)^{40}$

15.　　$\mu = (50)(0.4) = 20$,
　　　var $= (50)(0.4)(0.6) = 12$, $\sigma = \sqrt{12} = 3.464$

17.　　$\mu = (600)(0.52) = 312$,
　　　var $= (600)(0.52)(0.48) = 149.76$, $\sigma = \sqrt{149.76} = 12.238$

19.　　$\mu = (470)(0.08) = 37.6$,
　　　var $= (37.6)(0.92) = 34.592$, $\sigma = \sqrt{34.592} = 5.881$

21.　　$np = 50(.7) = 35 \geq 5$, $nq = (50)(.3) = 15 \geq 5$ so the normal
　　　distribution is a good estimate.

23.　　$np = (40)(.9) = 36$, $nq = (40)(.1) = 4 < 5$ so the normal
　　　distribution is not a good estimate.

25.　　$np = (25)(.5) = 12.5 \geq 5$, $nq = (25)(.5) = 12.5 \geq 5$ so the normal
　　　distribution is a good estimate.

27.　　$\mu = (50)(.7) = 35$, $\sigma = \sqrt{(35)(.3)} = \sqrt{10.5} = 3.24$
　　　(a)　　For $x_1 = 39.5$, $z_1 = (39.5 - 35)/3.24 = 1.39$
　　　　　　For $x_2 = 40.5$, $z_2 = (40.5 - 35)/3.24 = 1.70$
　　　　　　$P(x = 40) = A_2 - A_1 = 0.4554 - 0.4177 = 0.0377$
　　　(b)　　For $x_1 = 27.5$, $z_1 = (27.5 - 35)/3.24 -2.31$
　　　　　　For $x_2 = 28.5$, $z_2 = (28.5 - 35)/3.24 = -2.01$
　　　　　　$P(X = 28) = A_2 - A_1 = 0.4896 - 0.4778 = .0118$

160

(c)   For $x_1 = 31.5$, $z_1 = (31.5 - 35)/3.24 = -1.08$
         For $x_2 = 32.5$, $z_2 = (32.5 - 35)/3.24 = -.77$
         $P(X = 32) = A_2 - A_1 = .3599 - .2794 = .0805$

29. $\mu = 15(.4) = 6$, $\sigma = \sqrt{15(.4)(.6)} = \sqrt{3.6} = 1.90$
     (a)   For $x_1 = 4.5$, $z_1 = (4.5 - 6)/1.90 = -.79$, so $A_1 = .2852$
           For $x_2 = 7.5$, $z_2 = (7.5 - 6)/1.90 = .79$, so $A_2 = .2852$
           $P(4 < x < 8) = .2852 + .2852 = .5704$
     (b)   For $x_1 = 3.5$, $z_1 = (3.5 - 6)/1.90 = -1.32$, so $A_1 = .4066$
           For $x_2 = 8.5$, $z_2 = (8.5 - 6)/1.90 = 1.32$, so $A_2 = .4066$
           $P(4 \leq x \leq 8) = .4066 + .4066 = .8132$
     (c)   For $x_1 = 6.5$, $z_1 = (6.5 - 6)/1.90 = .26$, so $A_1 = .1026$
           For $x_2 = 8.5$, $z_2 = 1.32$, so $A_2 = .4066$
           $P(7 \leq X \leq 8) = .4066 - .1026 = .3040$

31. $\mu = 8$, $\sigma = 2$
     (a)   For $x = 5.5$, $z = (5.5 - 8)/2 = -1.25$, so $A = .3944$
           $P(x > 5) = .5000 + .3944 = .8944$
     (b)   For $x = 4.5$, $z = (4.5 - 8)/2 = -1.75$, so $A = .4599$
           $P(x \geq 5) = .5000 + .4599 = .9599$
     (c)   For $x = 9.5$, $z = (9.5 - 8)/2 = .75$, so $A = .2734$
           $P(x > 9) = .5000 - .2734 = .2266$

33. $\mu = 7.2$, $\sigma = 2.24$
     (a)   For $x = 9.5$, $z = (9.5 - 7.2)/2.24 = 1.03$, so $A = .3485$
           $P(x < 10) = .5000 + .3485 = .8485$
     (b)   For $x = 10.5$, $z = (10.5 - 7.2)/2.24 = 1.47$, so $A = .4292$
           $P(x \leq 10) = .5000 + .4292 = .9292$
     (c)   For $x = 5.5$, $z = (5.5 - 7.2)/2.24 = -.76$, so $A = .2764$
           $P(x < 6) = .5000 - .2764 = .2236$

35. $\mu = 60$, $\sigma = 4.90$
     (a)   For $x_1 = 49.5$, $z_1 = (49.5 - 60)/4.90 = -2.14$, so $A_1 = .4838$
           For $x_2 = 75.5$, $z_2 = (75.5 - 60)/4.90 = 3.16$, so $A_2 = .4992$
           $P(50 \leq x \leq 75) = .4838 + .4992 = .9830$
     (b)   For $x = 75.5$, $z = 3.16$, so $A = .4992$
           $P(x > 75) = .5000 - .4992 = .0008$
     (c)   For $x = 49.5$, $z = -2.14$, so $A = .4838$
           $P(x \leq 50) = .5000 - .4838 = .0162$

37. $p = 1/3$, find $P(6 \leq x \leq 8)$.
     $\mu = (20)(1/3) = 6.67$,
     $\sigma = \sqrt{(20)(1/3)(2/3)} = \sqrt{4.44} = 2.11$
     For $x_1 = 5.5$, $z_1 = (5.5 - 6.67)/2.11 = -0.55$, so $A_1 = .2088$
     For $x_2 = 8.5$, $z_2 = (8.5 - 6.67)/2.11 = 0.87$, so $A_2 = .3079$
     $P(6 \leq x \leq 8) = 0.2088 + 0.3079 = 0.5167$

39. $p = 1/2$, find $P(50 \leq x \leq 55)$. $\mu = 50$, $\sigma = \sqrt{25} = 5$
     For $x_1 = 49.5$, $z_1 = (49.5 - 50)/5 = -0.1$, so $A_1 = .0398$
     For $x_2 = 55.5$, $z_2 = (55.5 - 50)/5 = 1.1$, so $A_2 = .3643$
     $P(50 \leq x \leq 55) = 0.0398 + 0.3643 = 0.4041$

Section 7.8

41. $p = 1/4$, find $P(20 \le x \le 24)$.

   $\mu = (64)(1/4) = 16$, $\sigma = \sqrt{(16)(3/4)} = \sqrt{12} = 3.46$
   For $x_1 = 19.5$, $z_1 = (19.5 - 16)/3.46 = 1.01$, so $A_1 = .3438$
   For $x_2 = 24.5$, $z_2 = (24.5 - 16)/3.46 = 2.46$, so $A_2 = .4931$
   $P(20 \le x \le 24) = 0.4931 - 0.3438 = 0.1493$

43. Estimate the probability that $x \le 250$.

   $\mu = (270)(.9) = 243$, $\sigma = \sqrt{(243)(.1)} = \sqrt{24.3} = 4.93$
   For $x = 250.5$, $z = (250.5 - 243)/4.93 = 1.52$
   $P(x \le 250) = 0.5 + 0.4357 = 0.9357$

45. Find $P(x \ge 65)$

   $\mu = (250)(.3) = 75$, $\sigma = \sqrt{(75)(.7)} = \sqrt{52.5} = 7.25$
   $x = 64.5$, $z = (64.5 - 75)/7.25 = 1.45$, so $A = .4265$
   $P(x \ge 65) = 0.5 + 0.4265 = 0.9265$

47. Let x = number correct
   (a) $3x - 1(90 - x) \ge 98$, $4x \ge 188$, $x \ge 47$

   (b) $\mu = (90)(1/2) = 45$, $\sigma = \sqrt{(45)(1/2)} = \sqrt{22.5} = 4.74$
   For $x = 46.5$, $z = (46.5 - 45)/4.74 = 0.32$
   $P(x \ge 47) = 0.5 - 0.1255 = 0.3745$

Section 7.8

1.   (a)   S.E. $= \sqrt{\dfrac{.45(.55)}{50}} = .070$ (b)   S.E. $= \sqrt{\dfrac{.32(.68)}{100}} = .047$

   (c)   S.E. $= \sqrt{\dfrac{.20(.80)}{400}} = .020$

3. $\bar{p} = .30$, $n = 50$. For $c = .90$ $A = .45$ which corresponds to $z = 1.65$

   SE $= \sqrt{(.30)(.70)/50} = 0.0648$
   E $= (1.65)(0.0648) = 0.1069$
   $.30 \pm .1069$ gives the interval $0.1931 < x < 0.4069$

5. $\bar{p} = .30$, $n = 50$. For $c = .98$, $A = .49$ which corresponds to $z = 2.33$

   SE $= \sqrt{(.30)(.70)/50} = 0.0648$
   E $= (2.33)(0.0648) = 0.1510$
   $.30 \pm .1510$ gives the interval $0.1490 < x < 0.4510$

7. $\bar{p} = .5$, $n = 400$. For $c = .95$, $A = .475$ which corresponds to $z = 1.96$

   SE $= \sqrt{(.5)(.5)/400} = 0.025$
   E $= (1.96)(0.025) = 0.049$
   $.5 \pm .049$ gives the interval $0.451 < x < 0.549$

9.   (a) $\bar{p} = .45$, $n = 300$. For $c = .90$, $A = .45$ which corresponds to $z = 1.65$

   SE $= \sqrt{(.45)(.55)/300} = 0.0287$
   E $= (1.65)(0.0287) = 0.0473$
   $.45 \pm .0474$ gives the interval $0.4026 < x < 0.4974$

(b) $\overline{p} = .45$, $n = 300$, For $c = .95$, $A = .475$ which corresponds to $z = 1.96$
$SE = 0.0287$ $E = (1.96)(0.0287) = 0.0563$
$.45 \pm .0563$ gives the interval $0.3937 < x < 0.5063$

11. $\overline{p} = 243/300 = .81$, $n = 300$, For $c = .98$, $A = .490$ which corresponds to $z = 2.33$
$SE = \sqrt{(.81)(.19)/300} = 0.0226$
$E = (2.33)(0.0226) = 0.0528$
$.81 \pm .0528$ gives the interval $0.7572 < x < 0.8628$

13.   (a)   The estimate for $\overline{p}$ is $\dfrac{144}{195} = .738 = 73.8\%$.

(b) $SE = \sqrt{\dfrac{.738(.262)}{195}} = .0315$
For a 95% confidence interval $z = 1.96$.
The error bounds are
        $.738 + 1.96(.0315) = .7997$
and $.738 - 1.96(.0315) = .6763$
The 95% confidence interval is $.6763 < x < .7997$.

(c)   At the 99% confidence level $z = 2.58$.
The error bounds are
$.738 + 2.58(.0315) = .8193$
$.738 - 2.58(.0315) = .6567$
The 99% confidence interval is $.6567 < x < .8193$.

15.   (a) $S.E. = 1.96\sqrt{\dfrac{.27(.73)}{300}} = .0502$
Error bounds: $0.27 - 0.0502 = 0.2198$
$0.27 + 0.0502 = 0.3202$

(b) $S.E. = 1.96\sqrt{\dfrac{.27(.73)}{1000}} = .0275$
Error bounds: $0.27 - 0.0275 = 0.2425$
        $0.27 + 0.0275 = 0.2975$

(c)   The larger sample yields a smaller confidence interval.

17.   (a) $\overline{p} = .28$, $n = 200$, For $c = .95$ $A = .475$ which corresponds to $z = 1.96$
$SE = \sqrt{(.28)(.72)/200} = 0.0317$
$E = (1.96)(0.0317) = 0.0622$
$.28 \pm .0622$ gives the interval $0.2178 < x < 0.3422$

(b) $SE = \sqrt{(.28)(.72)/400} = 0.0224$
$E = (1.96)(0.0224) = 0.0439$
$.28 \pm .0439$ gives the interval $0.2361 < x < 0.3239$

19. $\overline{p} = 144/320 = .45$, $n = 320$, For $c = .95$, $A = .475$ which corresponds to $z = 1.96$
$SE = \sqrt{(.45)(.55)/320} = 0.0278$
$E = (1.96)(0.0278) = 0.0545$
$.45 \pm .0545$ gives the interval $0.3955 < x < 0.5045$

21. $\bar{p} = .54$, n = 60, For c = .90, A = .45 which corresponds to z = 1.65

$$SE = \sqrt{(.54)(.46)/60} = 0.0643$$
E = (1.65)(0.0643) = 0.1061
.54 $\pm$ .1061 gives the interval $0.4339 < x < 0.6461$

23. E = .06, c = .95 corresponds to z = 1.96 so

$.06 = 1.96\sqrt{.25/n}$
0.000937 = .25/n,
n = 266.8 rounded to 267

25. E = .02, c = .95 which corresponds to z = 1.96 so

$.02 = 1.96\sqrt{.25/n}$
0.000104123 = .25/n
n = 2401

27. In each case $\bar{p} = .65$ and z = 1.96.

(a) $SE = \sqrt{\dfrac{.65(.35)}{300}} = .0275$

The error bounds are
.65 + 1.96(.0275) = .7039
.65 - 1.96(.0275) = .5961

(b) $SE = \phantom{\sqrt{xx}} = .0251$

The error bounds are
.65 + 1.96(.0251) = .6992
.65 - 1.96(.0251) = .6008

(c) $SE = \sqrt{\dfrac{.65(.35)}{400}} = .0238$

The error bounds are
.65 + 1.96(.0238) = .6966
.65 - 1.96(0.238) = .6034

(d) $SE = \sqrt{\dfrac{.65(.35)}{460}} = .0222$

The error bounds are
.65 + 1.96(.0222) = .6935
.65 - 1.96(.0222) = .6065

(e) $SE = \sqrt{\dfrac{.65(.35)}{500}} = .0213$

The error bounds are
.65 + 1.96(.0213) = .6917
.65 - 1.96(.0213) = .6083

(f) $SE = \sqrt{\dfrac{.65(.35)}{1000}} = .0151$

The error bounds are
.65 + 1.96(.0151) = .6796
.65 - 1.96(.0151) = .6204

29. E = .0785, c = .95 which corresponds to z = 1.96 so

$.0785 = 1.96\sqrt{.25/n}$
0.001604 = .25/n
n = 155.8 rounded to 156

Section 7.8

31. $\bar{p}$ is the midpoint of 0.5007 and 0.5393, $\bar{p} = 0.52$.
    The upper bound is

    $$.52 + 1.65\sqrt{\frac{.52(.48)}{n}} \text{ where } z = 1.65$$

    corresponds to A = .45

    Thus, $.5393 = .52 + 1.65\sqrt{\dfrac{.2496}{n}}$

    $$\frac{.0193}{1.65} = \sqrt{\frac{.2496}{n}}$$

    $$n = .2496\left(\frac{1.65}{.0193}\right)^2 = 1824$$

    The sample contained 1824 students.

33. $p = .60$, $n = 300$, $SE = \sqrt{\dfrac{.60(.40)}{300}} = .02828$

    For $\bar{p} = 54\%$ $z = \dfrac{.54 - .60}{.02828} = -2.12$

    The area under the normal curve between the mean and z = −2.12 is
    .4830. The area less than z = −2.12 is .5000 − .4830 = .0170.
    The probability that less than 54% of the sample favor the fees is
    .017.

35. For all random samples of size 500, the proportions of out of
    state students in the samples form a normal distribution with the
    population proportion p = .45 as the mean. The probability a
    sample proportion is ≥ .50 is the fraction of the area above .50

    for the normal curve with mean .45 and $\sigma = \sqrt{\dfrac{.45(.55)}{500}}$

    $p = 0.45$, $n = 500$, $SE = \sqrt{\dfrac{.45(.55)}{500}} = .02225$

    For $\bar{p} = 0.50$ $z = \dfrac{.50 - .45}{.02225} = 2.25$

    A = .4878 corresponds to z = 2.25

    Thus the probability $\bar{p} > .50$ is .5000 − .4878 = .0122.
    The probability that at least 50% are from out-of-state is .0122.

37. $SE = \sqrt{\dfrac{.46(.54)}{1498}} = .012877$

    The margin of error ±03 comes from ±z × SE so
    $$.03 = z(.012877)$$
    $$z = 2.33$$

    z = 2.33 corresponds to A = .4901 since the confidence interval
    extends on both sides of $\bar{p}$, 2 × .4901 = .9802 = 98% gives the 98%
    confidence level.

39. The sample proportion of defective labels is
    $$\bar{p} = \frac{6}{1600} = .00375 = .3759\%$$

    $$SE = \sqrt{\frac{.00375(.99625)}{1600}} = .00153$$

    at the 95% confidence level the error bounds are
    $$.00375 + 1.96(.00153) = .00675$$

and $.00375 - 1.96(.00153) = .000751$

So the true proportion of defective labels is in the interval $.000751 < p < .00675$ with probability 0.95, or equivalently, between .0751% and .675%.
The manager is 95% confident that less than 0.7% of the labels are defective so the machine should not be shut down.

41. We are looking for $\bar{p}$ such that

$$\bar{p} = .68 + z\sqrt{\frac{.68(.32)}{4500}}$$

where z corresponds to an area $.5000 - .0808 = .4192$
(.0808 of the area is above z under the standard normal curve.)
For A = .4192 z = 1.40 so

$$\bar{p} = .68 + 1.40\sqrt{\frac{.68(.32)}{4500}} = .68 + 1.40(.006954) = .6897$$

The sample proportion would exceed .6897 only about 8.08% of the time.

Review Exercises, Chapter 7

1.

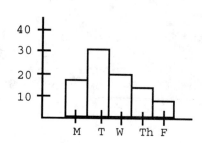
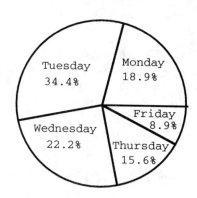

3.   (a)   6   (b) $(9 + 12)/2 = 21/2 = 10.5$
     (c) $(1 + 3)/2 = 2$

5. $\mu = \dfrac{54(1.5) + 32(3.5) + 12(5.5)}{98} = 259/98 = 2.64$

7. $90.22/n = 3.47$
    $n = 90.22/3.47 = 26$ 9. $X \in \{0, 1, 2, 3\}$

11. $E(X) = .05(0) + .20(1) + .15(2) + .20(3) + .25(4) + .15(5) = 2.85$

13. $E(X) = .20(1) + .32(2) + .21(3) + .15(4) + .12(5) = 2.67$

15.   For x = 400, $z = (400 - 350)/25 = 2$
    $.5 - A = .5 - 0.4773 = 0.0227$

17.

X	P(X)	
0	$C(4, 0)(.25)^0(.75)^4$	$= 0.3164$
1	$C(4, 1)(.25)^1(.75)^3$	$= 0.4219$
2	$C(4, 2)(.25)^2(.75)^2$	$= 0.2109$
3	$C(4, 3)(.25)^3(.75)^1$	$= 0.0469$
4	$C(4, 4)(.25)^4(.75)^0$	$= 0.0039$

19. $\mu = (80)(.65) = 52$, $\sigma = \sqrt{(52)(.35)} = \sqrt{18.2} = 4.266$
    (a)   For $x = 50.5$, $z = (50.5 - 52)/4.266 = -0.35$ and $A = .1368$
           $P(X > 50) = .5 + 0.1368 = 0.6368$
    (b)   For $x_1 = 64.5$, $z_1 = (64.5 - 52)/4.266 = 2.93$ and $A_1 = .4983$
           For $x_2 = 65.5$, $z_2 = (65.5 - 52)/4.266 = 3.16$ and $A_2 = .4992$
           $P(X = 65) = A_2 - A_1 = 0.4992 - 0.4983 = 0.0009$
    (c)   For $x_1 = 55.5$, $z_1 = (55.5 - 52)/4.266 = 0.82$ and $A_1 = .2939$
           For $x_2 = 59.5$, $z_2 = (59.5 - 52)/4.266 = 1.76$ and $A_2 = .4608$
           $P(55 < X < 60) = A_2 - A_1 = 0.4608 - 0.2939 = 0.1669$

21. $216 - 43 + 1 = 174$ was the number who placed the same or lower so
$174/216 = .805$ shows she was at the 81st percentile

23.

X	P(X)
1	20/800 = 0.025
2	160/800 = 0.200
3	370/800 = 0.4625
4	215/800 = 0.26875
5	35/800 = 0.04375

25.

X	Number of outcomes
0	10
1	30
2	15

27.

x	Number of ways
1	1
2	1
3	4

29. $\mu = \dfrac{8(3) + 13(8.5) + 6(12.5) + 15(17.5)}{42} = 11.24$

31.   (a) $\sqrt{\dfrac{(.35)(.65)}{60}} = .0616$ (b) $\sqrt{\dfrac{(.64)(.36)}{700}} = .0181$

    (c) $\sqrt{\dfrac{(.40)(.60)}{950}} = .0159$

33.   (a) $.733 \pm 1.96\sqrt{\dfrac{.733(.267)}{30}}$
        $.575 < p < .891$

Chapter 8
Mathematics of Finance

Section 8.1

1. $I = (1100)(.08)(9/12) = \$66$

3. $I = (600)(.10)(18/12) = \$90$

5. $I = (745)(.085)(6/12) = 31.6625 = \$31.66$

7. $I = (300)(.06)(1) = \$18 \quad A = P + I = 300 + 18 = \$318$

9. $I = (500)(.08)(3) = \$120$

11. $I = (700)(.015)(5) = \$52.50$

13. $I = (950)(.0175)(7) = 116.375 = \$116.38$

15. $P = I/rt = \dfrac{42.90}{(.13)(.5)} = \$660$

17. $P = I/rt = \dfrac{116.1}{(.09)(1.5)} = \$860$

19. $r = I/Pt = \dfrac{49.4}{(1140)(8/12)} = .065 = 6.5\%$

21. $r = I/Pt = \dfrac{616.25}{(5800)(1.25)} = .085 = 8.5\%$

23. $A = (2700)[1 + (.08)(1.5)] = \$3024$

25. $A = (6500)[1 + (0.086)(1.75)] = \$7478.25$

27. $P = \dfrac{1800}{1 + (.08)(1.5)} = 1607.142 = \$1607.14$

29. $D = (1850)(.075)(1) = \$138.75, \quad PR = M - D = \$1711.25$

31. $D = (485)(.13)(1.5) = \$94.58, \quad PR = \$390.42$

33. $I = 3500(.076)(2.5) = 665$
    Total interest paid was $665.

35. One year's simple interest is
    $I = 17500(.0685)(1) = \$1198.75$

37. (a) For one month $I = 85000(.081)(\frac{1}{12}) = 573.75$

    (b) Total interest payment
    $I = 85000(.081)(5) = 34,425$
    Total interest = $34,425

39. From $I = Prt, \quad t = I/Pr = 144/(800)(0.06) = 3$ years

41. From $I = Prt, \quad t = I/Pr = \dfrac{18}{(800)(.09)} = 0.25$ years $= 3$ months

43. $P = I/rt = \dfrac{48.75}{(0.075)(1)} = \$650$

45. $I = (20,000)(.09)(1/12) = \$150$ toward interest,
$179.95 - 150 = \$29.95$ toward principal

47.   In this case $P = 860$ and $A = 900$. Using $A = P(1 + rt)$,
$$900 = 860(1 + r(.5))$$
$$40 = 430r$$
$$r = .0930 \quad = 9.3\%$$

49. $I = 50$. From $I = Prt$, $50 = 950(r)(.5)$ $r = \dfrac{50}{475} = .10526 = 10.53\%$

51.   Simple interest: $I = (3000)(.103)(4/12) = \$103$

Simple discount: $M = \dfrac{PR}{1 - dt} = \dfrac{3000}{1 - (.101)(4/12)} = 3104.5188$ so
$D = \$104.52$. Simple interest at 10.3% results in the lower fee.

53. $I = (450)(.11)(1)$ billion $= \$49.5$ billion

55. $I = (500,000)(.09)(5) = \$225,000$

57.   (a)   Semiannual interest $= 625000(.085)(.5) = \$26,562.50$
(b)   Total interest $= 625000(.085)(8) = \$425,000$.

59. $PR = M(1 - dt) = (1 \text{ million})[1 - (.07)(90/360)] = \$982,500$

61. $982,000 = 1,000,000(1 - d(\dfrac{90}{360}))$
$-18000 = -250,000d$
$d = \dfrac{18,000}{250,000} = .072 = 7.2\%$

63. $PR = M(1 - dt) = (2 \text{ million})[1 - (.07125)(30/360)]$
$= \$1,988,125$

65.   Quarterly interest $= 350,000(.072)(\dfrac{1}{4}) = \$6300$
Total interest $= 6300(24) = \$151,200$

Section 8.2

1.   (a) $i = .15/2 = .075 = 7.5\%$ (b) $i = .15/4 = .0375 = 3.75\%$
(c) $i = .15/12 = .0125 = 1.25\%$

3.   (a) $A = 4500(1.1.075^6) = \$6,944.86$
(b) $I = A - P = \$2,444.86$

5.   (a) $A = 31,000(1.075^8) = \$55,287.81$ (b) $I = \$24,287.81$

7.   (a) $A = 5000(1.07^4) = \$6,553.98$ (b) $I = \$1,553.98$

9.   First quarter: $A = 800(1.03) = \$824$
Second quarter: $A = 800(1.03)^2 = \$848.72$
Third quarter: $A = 800(1.03)^3 = \$874.18$
Fourth quarter: $A = 800(1.03)^4 = \$900.41$

11. $A = 1800(1.045)^2 = \$1,965.65$

13.   (a) $A = (12,000)(1.10)^3 = \$15,972$

       (b) $A = (12,000)(1.05)^6 = \$16,081.15$

       (c) $A = (12,000)(1.025)^{12} = \$16,138.67$

15. $A = (10,000)(1.03)^{20} = \$18,061.11$

17. $A = (460)(1.008333)^6 = 460(1.051051) = \$483.48$

19. $A = (640)(1.02)^6 = \$720.74$

21. $A = (32.75)(1.01)^4 = \$34.08$

23. $P = \dfrac{25000}{(1.03)^{30}} = \$10,299.67$

25. $P = \dfrac{A}{(1 + i)^n} = \dfrac{3000}{(1.025)^{16}} = \$2,020.87$

27. $x = (1.03)^2 - 1 = .0609 = 6.09\%$

29.   By trial and error find n so that $2 = (1.03)^n$. $(1.03)^n$ reaches 2 during the 24th semiannual period so n = 24 semiannual periods = 12 years

31.   By trial and error find n so that $2 = (1.03)^n$. $(1.03)^n$ reaches 2 during the 24th quarter so n = 24 quarters = 6 years

33. $2 = (1 + i)^{28}$ The 28th root of 2, $2^{1/28}$, is 1.02506 so we can say 2.5% is the quarterly rate which gives 10% annual rate.

35.   By trial and error find n so that $3 = (1.05)^n$. $(1.05)^n$ reaches 3 during the 23rd semiannual period so n = 23 = 11.5 years

37. $A = 260(1.016)^5 = \$281.48$

39.   Ken had $1000(1.023)^{20} = \$1575.84$

      Barb had $1000(1.0076)^{60} = \$1575.03$ Ken had \$.81 more.

41.   The effective rate for 8.8% is $(1.044)^2 - 1 = .089936 = 8.9936\%$. The effective rate for 8.6% is $(1.0215)^4 - 1 = .088813 = 8.8813\%$, so 8.8% compounded semiannually is better.

43.   The effective rate of 9.4% compounded annually is 9.4%. The effective rate of 9.2% compounded quarterly is $(1.023)^4 - 1 = .09522 = 9.522\%$, so 9.2% is better.

45.   The effective rate of 10% compounded quarterly is $(1.025)^4 - 1 = .1038 = 10.38\%$, which is better than 10.2% compounded annually.

47. $A = (15,000)(1.02)^{28} = \$26,115.36$

49. $1485.95 = 1000(1 + r)^{20}$, so $(1 + r)^{20} = 1.48595$. $1 + r = 1.48595^{1/20}$ which is approximately 1.02, i = .02, so the annual rate is 8%.

51. Find i so that $633.39 = 500(1 + i)^8$ which is the same as $1.26678 = (1 + i)^8$ $1.26678^{1/8} = 1.030001$ which we round to 1.03 so $i = .03$ which gives 6% per year.

53. For $6000 invested for 10 years, $A = (6,000)(1.05)^{20} = \$15,919.79$ Take the lump sum of $16,000.

55. By trial and error find n so that $18,000 = 8,000(1.03)^n$ which is the same as $2.25 = (1.03)^n$. This occurs during the 28th semiannual period so n = 28 half years = 14 years.

57. $P = \dfrac{240,000}{(1.04)^{10}} = \$162,135.40$

59. (a) $20,000 is the future value and $10,250 is the present value so
$$20,000 = 10250(1 + i)^{10}$$
$1.95122 = (1 + i)^{10}$ so $\sqrt[10]{1.95122} = 1 + i$
$1.069130 = 1 + i$ so $i = .069130$ which we round to .0691.
i = 6.91%

   (b) $10000 = P(1.062)^6$
$P = \dfrac{10,000}{1.062^6} = \dfrac{10,000}{1.43465} = 6970.32$ She should pay $6970.32.

67. For 5% interest

Years	Amount
50	$ 52,750.04
100	$ 604,905.79
150	$ 6,936,696.48
200	$79,545,871.75

For 4.5% after 200 years A = $30,620,756.86.
For 5.5% after 200 years A = $3,205,707,325.80

73. (a) $162.89   (b) 6,728,970

Section 8.3

1.

Year Deposited	Value at End of 4 Years
1	$600(1.1)^3 = \$~798.60$
2	$600(1.1)^2 = \$~726.00$
3	$600(1.1)^1 = \$~660.00$
4	$600(1.1)^0 = \$~600.00$
	$2,784.60 Final Value

3. n = 15, i = .07
$$A = 16,000\left[\dfrac{1.07^{15} - 1}{.07}\right] = \$402,064.35$$

5. n = 20, i = .08/4 = .02 $A = 250\left[\dfrac{1.02^{20} - 1}{.02}\right] = \$6074.34$

7. $n = 20$, $i = .01$ $\qquad A = 200\left[\dfrac{1.01^{20} - 1}{.01}\right] = \$4403.80$

9. $n = 20$, $i = .04$ $\qquad A = 4000\left[\dfrac{1.04^{20} - 1}{.04}\right] = \$119{,}112.31$

11. $n = 16$, $i = .12/4 = .03$ $\qquad A = 750\left[\dfrac{1.03^{16} - 1}{.03}\right] = \$15{,}117.66$

13. $2500 = R\left[\dfrac{1.09^{4} - 1}{.09}\right]$ $\qquad R = \$546.67$

15. $14{,}500 = R\left[\dfrac{1.05^{20} - 1}{.05}\right]$ $\qquad R = \$438.52$

17. $10{,}000 = R\left[\dfrac{1.025^{12} - 1}{.025}\right]$ $R = \$724.87$

19. $75{,}000 = R\left[\dfrac{1.025^{16} - 1}{.025}\right]$ $R = \$3869.92$

21. $15{,}000 = R\left[\dfrac{1.01^{18} - 1}{.01}\right]$ $\qquad R = \$764.73$

23. $n = 24$, $i = .01$ $\qquad A = 100\left[\dfrac{1.01^{24} - 1}{.01}\right] = \$2697.35$

25. $n = 10$, $i = .04$ $\qquad A = 400\left[\dfrac{1.04^{10} - 1}{.04}\right] = \$4{,}802.44$

27. $n = 12$, $i = .015$ $\qquad 4000 = R\left[\dfrac{1.015^{12} - 1}{.015}\right]$ $R = \$306.72$

29. $n = 9$, $i = .07$ $\qquad A = 5{,}000\left[\dfrac{1.07^{9} - 1}{.07}\right] = \$59{,}889.94$

31. $n = 6$, $i = .07$ $\qquad 150{,}000 = R\left[\dfrac{1.07^{6} - 1}{.07}\right]$ $R = \$20{,}969.37$

33. $n = 16$, $i = .04$ $\qquad 750{,}000 = R\left[\dfrac{1.04^{16} - 1}{.04}\right]$ $R = \$34{,}365.00$

35. $n = 5$, $i = .06$, $R = 4{,}000$,

The amount accumulated after 5 years $= 4000\left[\dfrac{1.06^{5} - 1}{.06}\right] = 22{,}548.37.$
This is less than the required amount.

37.   (a)   R = 1000, n = 10, i = .08,

$$A = 1000\left[\frac{1.08^{10} - 1}{.08}\right] = \$14,486.56$$

(b)   .08(14,486.56) = \$1158.92   The annual interest of \$1158.92 will make the \$1000 payment.

39. $A = 100\left[\frac{(1.016)^{30} - 1}{.016}\right] = \$3812.16$

41. $15,000 = R\left[\frac{(1.0075)^{60} - 1}{.0075}\right] = R(75.424137)$

R = \$198.88

43. $100,000 = R\left[\frac{(1.081)^{6} - 1}{.081}\right] = R(7.3544534)$

R = \$13,597.20

45.   i = .07/12 = .00583333
    (a)   n = 300, A = \$162,014   (b)   n = 312, A = \$176,205
    (c)   n = 324, A = \$191,421   (d)   n = 336, A = \$207,738
    (e)   n = 348, A = \$225,234   (f)   n = 360, A = \$243,994

47.   i = .06
    (a)   n = 40, A = \$309,524   (b)   n = 35, A = \$222,870
    (c)   n = 25, A = \$109,729   (d)   n = 15, A = \$ 46,552

51.   x = .006238 per month = 7.49% annual

Section   8.4

1. $n = 8, i = .07, P = 4000\left[\frac{1 - (1.07)^{-8}}{.07}\right] = \$23,885.19$

3.   R = 508.80, i = .0075, and n = 36

$$P = 508.8\left[\frac{1 - (1.0075)^{-36}}{.0075}\right] = 16,000.13$$

The present value is \$16,000.

5.   R = 240, i = .007, n = 120.

$$P = 240\left[\frac{1 - (1.007)^{-120}}{.007}\right] = 19440.83$$

The present value is \$19,440.83.

7. $n = 22, i = .025, P = 300\left[\frac{1 - (1.025)^{-22}}{.025}\right] = \$5029.62$

9.   n = 12, i = .02

$$P = 226.94\left[\frac{1 - (1.02)^{-12}}{.02}\right] = \$2,399.97 \text{ rounded to } \$2,400$$

173

11. $R = 395.17$, $i = .0075$, $n = 24$

$$P = 395.17\left[\frac{1 - (1.0075)^{-24}}{.0075}\right] = 8649.93 \quad \text{The amount borrowed was \$8650.}$$

13. $$9500 = R\left[\frac{1-(1.069)^{-5}}{.069}\right] = R\left[\frac{1.069^5 - 1}{.069(1.069)^5}\right]$$

$9500 = R(4.111199)$

$R = 2310.76$ The annual payments are \$2310.76.

15. $$7500 = R\left[\frac{1 - (1.018)^{-60}}{.018}\right]$$

$7500 = R(36.507054)$

$R = \$205.44$ The quarterly payments are \$205.44.

17. $$32000 = R\left[\frac{1 - (1.006)^{-96}}{.006}\right]$$

$32000 = R(72.814868)$

$R = 439.47$ The monthly payments are \$439.47.

19. $$96000 = R\left[\frac{1 - (1.00475)^{-180}}{.00475}\right]$$

$96000 = R(120.811594)$

$R = 794.626$ The monthly payments are \$794.63.

21. $i = .102/12 = .0085$, $n = 48$

$$12750 = R\left[\frac{1 - (1.0085)^{-48}}{.0085}\right]$$

$12750 = R(39.2792)$

$R = 324.5989$ The monthly payments are \$324.60.

23. $i = .096/12 = .008$, $n = 360$

$$68000 = R\left[\frac{1 - (1.008)^{-360}}{.008}\right]$$

$68000 = R(117.90229)$

$R = 576.7488$ The monthly payments are \$576.75.

25. (a) $(360)(825.75) = \$297,270$ (b) $297,270 - 75,000 = \$222,270$

27. (a) $I = (68,000)(.15)(1/12) = \$850$ interest
    $870.96 - 850 = \$20.96$ to principal
    (b)   Total paid $= (25)(12)(870.96) = \$261,288$

29. $n = 20$, $i = .025$

$$\text{Bal} = (75,000)(1.025)^{20} - (2987.72)\left[\frac{1.025^{20} - 1}{.025}\right]$$

$$= 122,896.23 - 76,320.28 = \$46,575.95$$

31. P = 2500, R = 150, n = 18

$$2500 = 150\left[\frac{1 - (1 + i)^{-18}}{i}\right]$$

$$16.667 = \frac{1 - (1 + i)^{-18}}{i}$$

By trial and error we find i = .00823 yields

$$16.667 \text{ for } \frac{1 - (1 + i)^{-18}}{i}$$

so the monthly interest rate is about 0.823% and the annual interest rate (APR) is about 9.876%.

33. The periodic monthly rate is .108/12 = .009 and n = 24

$$P = 400.03\left[\frac{1 - 1.009^{-24}}{.009}\right] = 8600.0144 \quad \text{She borrowed \$8600.}$$

35. n = 16, i = .025

$$7500 = R\left[\frac{1 - 1.025^{-16}}{.025}\right] \quad R = \$574.49$$

37. n = 20, i = .09

$$P = 10,000\left[\frac{1 - 1.09^{-20}}{.09}\right] = \$91,285.46$$

39. n = 24, i = .01

$$12,000 = R\left[\frac{1 - 1.01^{-24}}{.01}\right] \quad R = \$564.88$$

41. (i) Find amount needed at age 18 to pay \$15,000 per year for 4 years.

n = 4, i = .08

$$P = (15,000)\left[\frac{1 - 1.08^{-4}}{.08}\right] = \$49,681.90 \text{ needed at 18}$$

(ii) Find periodic payments of an annuity that will accumulate to \$49,681.90 in 18 years.

n = 18, i = .08

$$49,681.90 = R\left[\frac{1.08^{18} - 1}{.08}\right]$$

R = \$1,326.61

43. $$P = 200\left[\frac{(1.0075)^{60} - 1}{.0075(1.0075)^{60}}\right] = 200(48.173374) = \$9634.67$$

45. $$9700 = R\left[\frac{(1.0075)^{60} - 1}{.0075(1.0075)^{60}}\right] = \$201.36$$

47. $$85,000 = R\left[\frac{(1.0075)^{120} - 1}{.0075(1.0075)^{120}}\right] = \$1076.74$$

49. $i = .0075$, $n = 300$ months

Monthly payments $= \dfrac{8000(.0075)}{1 - 1.0075^{-300}} = \$671.36$

Total payments over two years $= 24(671.36) = \$16,112.64$

Balance after two years $= 80000(1.0075)^{24} - 671.36\left[\dfrac{1.0075^{24} - 1}{.0075}\right]$

$$= \$78,131.19$$

Costs of owning the house

Down payment	$10,000.00
Payment of balance	78,131.19
Monthly payments	16,112.64
Real estate agent	5,400.00
Closing costs	1,350.00
	110,993.83
Income from sale of house	90,000.00
Net cost	20,993.83

Average monthly cost $= 20,993.83/24 = \$874.74$
Note: This does not include the cost of taxes, insurance, and utilities.

55. First, find the monthly payments from

$$75,000 = R\left[\dfrac{1 - (1.00666667)^{-360}}{.00666667}\right]$$

$R = \$550.32$

(a) The first month's interest on \$75,000 is \$500 so \$50.32 is repaid the first month.

(b) We need to find the balance after 180 months. It is

$$\text{Bal} = 75,000(1.00666667)^{180} - \dfrac{550.32[(1.00666667)^{180} - 1]}{.00666667}$$

$$= 248,019.26 - 190,431.82 = 57,587.44$$

The monthly interest on \$57,587.44 is \$383.92 so the principal repaid is \$550.32 - \$383.92 = \$166.40.

61. Graph $y = 500\left[\dfrac{(1 + x)^5 - 1}{x}\right]$ and trace until $y = 2886.87$.

The value of x that corresponds to this is $x = .072$.
The annual rate is 7.2%.
Trace the graph to determine the value of x that corresponds to $y = 1800$. We find $x = .0113$ or 1.13% per month which gives an annual rate of 13.56%.

69. $x = .01515$ per month $= 18.18\%$ annual

71. $x = .00616$ per month $= 7.39\%$ annual

73. (a) $R = 50,000$, $I = 3\%$, $r = 7\%$, $n = 15$

$$P = 50000\left(\dfrac{1.03}{.04}\right)\left[1 - \left(\dfrac{1.03}{1.07}\right)^{15}\right] = 560,476$$

(b) $R = 50,000$, $i = 7\%$, $n = 15$ for the present value of an annuity

$$P = 50000\left[\dfrac{1 - (1.07)^{-15}}{.07}\right] = 455,396$$

(c) $R = 35,000$, $I = 5\%$, $r = 8\%$, $n = 10$

$$P = 35000 \left(\frac{1.05}{.03}\right) \left[1 - \left(\frac{1.05}{1.08}\right)^{10}\right] = 300,746$$

(d) $R = 35,000$, $I = 10\%$, $r = 12\%$, $n = 10$

$$P = 35000 \left(\frac{1.10}{.02}\right) \left[1 - \left(\frac{1.10}{1.12}\right)^{10}\right] = 317,402$$

(e) $R = 35,000$, $I = 1\%$, $r = 4\%$, $n = 10$

$$P = 35000 \left(\frac{1.01}{.03}\right) \left[1 - \left(\frac{1.01}{1.04}\right)^{10}\right] = 299,010$$

(f) $R = 35,000$, $I = 10\%$, $r = 13\%$, $n = 10$

$$P = 35,000 \left(\frac{1.10}{.03}\right) \left[1 - \left(\frac{1.10}{1.13}\right)^{10}\right] = 302,756$$

(g)   Part (e) with the lowest inflation and interest rates requires the smallest initial fund amount and part (f) with high inflation and interest rates requires the largest initial amount so it appears that low inflation and interest rates are preferable.

(h) $R = 35,000$, $I = 0$, $r = 3\%$, $n = 10$

$$P = 35000 \left(\frac{1}{.03}\right) \left[1 - \left(\frac{1}{1.03}\right)^{10}\right] = 298,557$$

This requires the smallest present value.

Review Exercises, Chapter 8

1. $I = (500)(.09)(2) = \$90$

3. $P = \dfrac{A}{1 + rt} = \dfrac{1190.40}{1 + (.08)(3)} = \$960$

5. $I = (3000)(.09)(5) = \$1,350$

7. $D = (8500)(.09)(2) = \$1,530$, $PR = 8500 - 1530 = \$6970$

9. $A = (5000)(1.07)^3 = 6125.21$ so interest $= \$1,125.21$

11.   By trial and error find n such that $2 = (1.02)^n$. This occurs during the 36th quarter so $n = 36$ quarters $= 9$ years

13. $x = (1.03)^2 - 1 = .0609 = 6.09\%$

15. $x = (1.02)^{12} - 1 = .26824 = 26.82\%$

17.   Effective rate of 18% $= (1.045)^4 - 1 = 0.1925 = 19.25\%$. Effective rate of 19% $= (1.19)^1 - 1 = 0.19$, so 18% compounded quarterly is better.

19.   (a) $A = (8,000)(1.06)^4 = \$10,099.82$
       (b) $I = 10099.82 - 8000 = \$2099.82$

21.   (a) $A = (5,000)(1.025)^{24} = \$9,043.63$
       (b) $I = 9043.63 - 5000 = \$4,043.63$

23. $A = (15,000)(1.08)^2 = \$17,496$

25. $50,000 = P(1.02)^{20} = \$33,648.57$

27. By trial and error find n such that $3500 = 2000(1.03)^n$ which is the same as $1.75 = (1.03)^n$. This occurs during the 19th quarter so n = 19 quarters = 4 years, 9 months

29. $A = 1,000 \left[\dfrac{1.06^5 - 1}{.06} \right] = \$5,637.09$

31. n = 10, i = .05 $\qquad A = 600 \left[\dfrac{1.05^{10} - 1}{.05} \right] = \$7,546.74$

33. n = 24, i = .02 $\qquad A = 250 \left[\dfrac{1.02^{24} - 1}{.02} \right] = \$7,605.47$

35. n = 10, i = .08

$\qquad 2,000,000 = R \left[\dfrac{1.08^{10} - 1}{.08} \right]$

$\qquad R = \$138,058.98$

37. $P = 5000/(1.06)^5 = \$3,736.29$

39. $P = 6000/(1.02)^{20} = \$4,037.83$

41. $P = 50,000/(1.02)^{20} = \$33,648.57$

43. By trial and error find n so that $297,000 = 200,000(1.02)^n$. This is the same as $1.485 = (1.02)^n$. The interest will exceed \$200,000 in the 20th quarter so the term should be 19 quarters, 4.75 years.

45. n = 12, i = .03

$\qquad 3000 = R \left[\dfrac{1 - 1.03^{-12}}{.03} \right] \qquad R = \$301.39$

47. n = 5, i = .09

49. n = 20, i = .07

$\qquad 7.8 \text{ million} = R \left[\dfrac{1 - 1.07^{-20}}{.07} \right] \qquad R = \$736,264.82$

51. n = 8, i = .09

$\qquad 98,000 = R \left[\dfrac{1 - 1.09^{-8}}{.09} \right] \qquad R = \$17,706.09$

53. Amount at end of first five years
$\quad A = 1000(1.02)^{20} = \$1485.95$
Amount at end of second five years
$\quad A = \$1485.95(1.025)^{20} = \$2434.90$

55. $A = 1700(1.0195)^{40} = \$3680.77$

57. $500,000 = R(1.025)^{160} = \$9619.48$

59. $100,000 = R \left[\dfrac{(1.01)^{72} - 1}{.01} \right] = R(104.70993) \quad R = \$955.02$

Chapter 9
Game Theory

Section 9.2

1.

$$\begin{array}{c} & & C \\ & & \begin{array}{cc} H & T \end{array} \\ R & \begin{array}{c} H \\ T \end{array} & \left[\begin{array}{cc} -1 & -.5 \\ -.5 & 2 \end{array} \right] \end{array}$$

3.

$$\begin{array}{c} & & C \\ & & \begin{array}{cc} 1 & 2 \end{array} \\ R & \begin{array}{c} 1 \\ 2 \end{array} & \left[\begin{array}{cc} 2 & -3 \\ -3 & 4 \end{array} \right] \end{array}$$

5. (a) R receives 10 from C   (b) 10   (c) 10   (d) 8

7.

Row Min.

$$\left[\begin{array}{cc} -3 & -5 \\ 2 & -1 \end{array} \right] \begin{array}{c} -5 \\ \boxed{-1} \end{array}$$

Col. Max. 2 $\boxed{-1}$

Strictly determined, the saddle point is (2, 2), value = -1, and the solution is row 2 and column 2.

9.

Row Min.

$$\left[\begin{array}{ccc} 1 & 2 & 3 \\ 4 & -5 & 1 \\ 2 & 6 & 3 \end{array} \right] \begin{array}{c} 1 \\ -5 \\ \boxed{2} \end{array}$$

Col. Max. 4 6 $\boxed{3}$

Not strictly determined

11.

Row Min.

$$\left[\begin{array}{ccc} 2 & 0 & 1 \\ 0 & -3 & 4 \\ 3 & -2 & 0 \end{array} \right] \begin{array}{c} \boxed{0} \\ -3 \\ -2 \end{array}$$

Col. Max.   3 $\boxed{0}$ 4

Strictly determined, the saddle point is (1, 2), value = 0, and the solution is row 1 and column 2.

13. (a)

Row Min.

$$\left[\begin{array}{ccc} -3 & 2 & 1 \\ 0 & 2 & 3 \\ -1 & -4 & 2 \end{array} \right] \begin{array}{c} -3 \\ \boxed{0} \\ -4 \end{array}$$

Col. Max. $\boxed{0}$ 2 3

Strictly determined, saddle point (2, 1), value = 0, solution row 2, column 1.

(b)

Row Min.

$$\left[\begin{array}{ccc} -1 & 2 & 3 \\ 4 & -1 & 0 \\ 0 & 1 & -1 \end{array} \right] \begin{array}{c} \boxed{-1} \\ \boxed{-1} \\ \boxed{-1} \end{array} \text{Max.}$$

Col. Max. 4 $\boxed{2}$ 3

Min.

Not strictly determined because the largest row minimum does not equal the smallest column maximum.

(c)

$$\begin{bmatrix} 1 & -2 & 1 \\ 5 & 7 & 3 \\ -1 & 3 & -4 \end{bmatrix} \begin{matrix} -2 \\ \boxed{3} \text{ Max.} \\ -4 \end{matrix}$$

Row Min.

Col. Max. 5 7 ③
Min.

Strictly determined, saddle point (2, 3), value = 3, solution row 2 and column 3.

(d)

$$\begin{bmatrix} -1 & -2 & 3 \\ 5 & 0 & 2 \\ 4 & -1 & 1 \end{bmatrix} \begin{matrix} -2 \\ \boxed{0} \text{ Max.} \\ -1 \end{matrix}$$

Row Min.

Col. Max. 5 ⓪ 3
Min.

Strictly determined, saddle point (2, 2), value = 0, solution row 2, column 2.

15.

$$\begin{bmatrix} 10 & -5 & 25 \\ -20 & -15 & 10 \\ -10 & -10 & 5 \end{bmatrix} \begin{matrix} \boxed{-5} \text{ Max.} \\ -20 \\ -10 \end{matrix}$$

Row Min.

Col. Max. 10 ⟨-5⟩ 25
Min.

ReMark should choose option 1, the sports highlights, and Century should choose option 2, the game show. The value is -5, ReMark should expect to lose 5 points in the ratings.

17. (a)

$$\begin{bmatrix} 85 & 120 & 150 \\ 60 & 165 & 235 \\ 70 & 150 & 175 \end{bmatrix} \begin{matrix} \boxed{85} \text{ Max.} \\ 60 \\ 70 \end{matrix}$$

Row Min.

Col. Max. ⟨85⟩ 165 235
Min.

The saddle point is (1, 1) so the farmer should plant milo.

(c) Total income over five years
Milo: 85 + 3(120) + 150 = 595
Corn: 60 + 3(165) + 235 = 790
Wheat: 70 + 3(150) + 175 = 695
In the long term corn appears to be the best and wheat second.

9.3 Exercises

1. $E = PAQ = \begin{bmatrix} \frac{2}{3} & \frac{1}{3} \end{bmatrix} \begin{bmatrix} 20 & 12 \\ 8 & 30 \end{bmatrix} \begin{bmatrix} \frac{1}{4} \\ \frac{3}{4} \end{bmatrix} = \begin{bmatrix} 16 & 18 \end{bmatrix} \begin{bmatrix} \frac{1}{4} \\ \frac{3}{4} \end{bmatrix} = 17.5$

3. $E = \begin{bmatrix} \frac{2}{5} & \frac{3}{5} \end{bmatrix} \begin{bmatrix} 3 & -6 \\ -2 & 4 \end{bmatrix} \begin{bmatrix} \frac{2}{3} \\ \frac{1}{3} \end{bmatrix} = \begin{bmatrix} 0 & 0 \end{bmatrix} \begin{bmatrix} \frac{2}{3} \\ \frac{1}{3} \end{bmatrix} = 0$

5. (a) $\begin{bmatrix} 1 & 0 & 0 \end{bmatrix} \begin{bmatrix} -15 & 40 & 25 \\ 25 & -10 & -5 \\ 45 & 20 & -15 \end{bmatrix} \begin{bmatrix} 0 \\ 1 \\ 0 \end{bmatrix} = 40$

   (b) $\begin{bmatrix} \frac{1}{2} & \frac{1}{2} & 0 \end{bmatrix} \begin{bmatrix} -15 & 40 & 25 \\ 25 & -10 & -5 \\ 45 & 20 & -15 \end{bmatrix} \begin{bmatrix} \frac{1}{2} \\ 0 \\ \frac{1}{2} \end{bmatrix} = \begin{bmatrix} 5 & 15 & 10 \end{bmatrix} \begin{bmatrix} \frac{1}{2} \\ 0 \\ \frac{1}{2} \end{bmatrix} = 7.5$

   (c) $\begin{bmatrix} \frac{1}{5} & \frac{2}{5} & \frac{2}{5} \end{bmatrix} \begin{bmatrix} -15 & 40 & 25 \\ 25 & -10 & -5 \\ 45 & 20 & -15 \end{bmatrix} \begin{bmatrix} \frac{1}{3} \\ \frac{1}{3} \\ \frac{1}{3} \end{bmatrix} = \begin{bmatrix} 25 & 12 & -3 \end{bmatrix} \begin{bmatrix} \frac{1}{3} \\ \frac{1}{3} \\ \frac{1}{3} \end{bmatrix} = \frac{34}{3}$

   (d) $\begin{bmatrix} .3 & .1 & .6 \end{bmatrix} \begin{bmatrix} -15 & 40 & 25 \\ 25 & -10 & -5 \\ 45 & 20 & -15 \end{bmatrix} \begin{bmatrix} .2 \\ .2 \\ .6 \end{bmatrix} = \begin{bmatrix} 25 & 23 & -2 \end{bmatrix} \begin{bmatrix} .2 \\ .2 \\ .6 \end{bmatrix} = 8.4$

7. $p_1 = \dfrac{20 - (-5)}{15 + 20 - 10 - (-5)} = \dfrac{25}{30} = \dfrac{5}{6}$ and $p_2 = \dfrac{1}{6}$

   $q_1 = \dfrac{20 - 10}{30} = \dfrac{1}{3}$ and $q_2 = \dfrac{2}{3}$

   Row strategy $= \begin{bmatrix} \frac{5}{6} & \frac{1}{6} \end{bmatrix}$ and column strategy $= \begin{bmatrix} \frac{1}{3} & \frac{2}{3} \end{bmatrix}$

   $E = \dfrac{15(20) - 10(-5)}{30} = \dfrac{350}{30} = 11.67$

9. Row 1 dominates row 2 and column 1 dominates column 2 so we can remove row 2 and column 2 from the matrix to obtain $\begin{bmatrix} 6 & 3 \\ 5 & 8 \end{bmatrix}$.
   We then have the optimal strategies
   $P = \begin{bmatrix} p_1 & 0 & p_3 \end{bmatrix}$ and $Q = \begin{bmatrix} q_1 & 0 & q_3 \end{bmatrix}$ where

   $p_1 = \dfrac{8 - 5}{6 + 8 - 3 - 5} = \dfrac{3}{6}$ and $p_3 = \dfrac{1}{2}$ so $P = \begin{bmatrix} \frac{1}{2} & 0 & \frac{1}{2} \end{bmatrix}$

   $q_1 = \dfrac{8 - 3}{6} = \dfrac{5}{6}$ and $q_3 = \dfrac{1}{6}$ so $Q = \begin{bmatrix} \frac{5}{6} & 0 & \frac{1}{6} \end{bmatrix}$

   $E = \dfrac{6(8) - 3(5)}{6} = \dfrac{33}{6} = \dfrac{11}{2}$

11. (a) Row 1 and row 2 dominate row 3, and column 2 dominates columns 1, 3, and 4, so row 3 and columns 1, 3 and 4 can be deleted leaving $\begin{bmatrix} -5 \\ 0 \end{bmatrix}$

(b) Column 1 dominates column 4, so the matrix reduces

to $\begin{bmatrix} 5 & 2 & -2 \\ 3 & 1 & 13 \\ 1 & 3 & 6 \end{bmatrix}$

13. (a) $E = \dfrac{6(4) - 3(8)}{6 + 4 + 3 + 8} = \dfrac{0}{21} = 0$. This is a fair game.

(b) $E = \dfrac{5(3) - 1(2)}{5 + 3 - 2 - 1} = \dfrac{13}{5}$. This is not a fair game.

(c) $E = \dfrac{3(6) - 2(-9)}{3 + 6 + 9 - 2} = \dfrac{36}{16}$. This is not a fair game.

15. $\begin{bmatrix} 10 & -5 & 25 \\ -20 & 15 & 10 \\ -10 & -10 & 5 \end{bmatrix}$ Since each entry in Row 1 is greater than the corresponding entry in Row 3, we can remove Row 3 from consideration. Since Column 1 dominates Column 3, we can remove Column 3 from consideration. We then have the reduced payoff matrix

$$R \begin{array}{c} C \\ \begin{bmatrix} 10 & -5 \\ -20 & 15 \end{bmatrix} \end{array}$$

For ReMark $p_1 = \dfrac{15 - (-20)}{10 + 15 - (-5) - (-20)} = \dfrac{35}{50} = \dfrac{7}{10}$,

$p_2 = \dfrac{3}{10}$ and $P = \begin{bmatrix} \dfrac{7}{10} & \dfrac{3}{10} & 0 \end{bmatrix}$.

For Century $q_1 = \dfrac{15 - (-5)}{50} = \dfrac{20}{50} = \dfrac{2}{5}$, $q_2 = \dfrac{3}{5}$, and $Q = \begin{bmatrix} \dfrac{2}{5} & \dfrac{3}{5} & 0 \end{bmatrix}$

$E = \dfrac{10(15) - (-5)(-20)}{50} = \dfrac{50}{50} = 1$

ReMark should show the sports highlights 70% of the time, the mystery drama 30% of the time, and drop the variety show. Century should air the talk show 40% of the time, the game show 60% of the time and drop the educational documentary.

Using these strategies ReMark can expect an average gain of 1 rating point.

17. (a) Expected survival time $= \begin{bmatrix} 1 & 0 \end{bmatrix} \begin{bmatrix} 25 & 30 \\ 5 & 35 \end{bmatrix} \begin{bmatrix} .60 \\ .40 \end{bmatrix}$

$= \begin{bmatrix} 25 & 30 \end{bmatrix} \begin{bmatrix} .60 \\ .40 \end{bmatrix} = 27$ years

(b) Expected survival time $= \begin{bmatrix} 0 & 1 \end{bmatrix} \begin{bmatrix} 25 & 30 \\ 5 & 35 \end{bmatrix} \begin{bmatrix} .60 \\ .40 \end{bmatrix}$

$= \begin{bmatrix} 5 & 35 \end{bmatrix} \begin{bmatrix} .60 \\ .40 \end{bmatrix} = 17$ years

(c)   No surgery is the better option when expected survival time
with no surgery is greater than expected survival time with
surgery.

$$[0 \quad 1] \begin{bmatrix} 25 & 30 \\ 5 & 35 \end{bmatrix} \begin{bmatrix} q_1 \\ 1 - q_1 \end{bmatrix} > [1 \quad 0] \begin{bmatrix} 25 & 30 \\ 5 & 35 \end{bmatrix} \begin{bmatrix} q_1 \\ 1 - q_1 \end{bmatrix}$$

$$[5 \quad 35] \begin{bmatrix} q_1 \\ 1 - q_1 \end{bmatrix} > [25 \quad 30] \begin{bmatrix} q_1 \\ 1 - q_1 \end{bmatrix}$$

$$5q_1 + 35 - 35q_1 > 25q_1 + 30 - 30q_1$$
$$35 - 30q_1 > 30 - 5q_1$$
$$5 > 25q_1$$
$$\frac{5}{25} > q_1$$

No surgery is the better option when $q_1 < .20$.

19.   (a)   Expected survival $= [1 \quad 0] \begin{bmatrix} 20 & 22 \\ 3 & 25 \end{bmatrix} \begin{bmatrix} .70 \\ .30 \end{bmatrix} = 20.6$ years

(b)   Expected survival $= [0 \quad 1] \begin{bmatrix} 20 & 22 \\ 3 & 25 \end{bmatrix} \begin{bmatrix} .70 \\ .30 \end{bmatrix} = 9.6$ years

(c)   Surgery is the better options when

$$[1 \quad 0] \begin{bmatrix} 20 & 22 \\ 3 & 25 \end{bmatrix} \begin{bmatrix} q_1 \\ 1 - q_1 \end{bmatrix} > [0 \quad 1] \begin{bmatrix} 20 & 22 \\ 3 & 25 \end{bmatrix} \begin{bmatrix} q_1 \\ 1 - q_1 \end{bmatrix}$$

$$20q_1 + 22(1 - q_1) > 3q_1 + 25(1 - q_1)$$
$$22 - 2q_1 > 25 - 22q_1$$
$$20q_1 > 3$$
$$q_1 > 3/20 = .15$$

Surgery is the better option when the probability of
malignancy is greater than .15.

Review Exercises, Chapter 9

1.

(a) $\begin{bmatrix} 5 & -1 \\ 2 & 4 \end{bmatrix} \begin{array}{c} -1 \\ 2 \end{array}$ This game is not strictly determined.
$\qquad\quad 5 \quad\; 4$

(b) $\begin{bmatrix} 1 & 3 & 9 \\ 7 & 4 & 8 \\ -5 & 3 & 4 \end{bmatrix} \begin{array}{c} 1 \\ 4 \\ 5 \end{array}$ This game is strictly determined.
$\qquad\quad 7 \quad 4 \quad 9$ The (2, 2) location is the saddle
point.   The value of the game is 4.

(c) $\begin{bmatrix} 140 & 210 \\ 300 & 275 \end{bmatrix} \begin{array}{c} 140 \\ 275 \end{array}$ This game is strictly determined.
$\qquad 300 \quad 275$ The (2, 2) location is the saddle
point.   The value of the game is 275.

(d) $\begin{bmatrix} -6 & 2 & 9 & 1 \\ 5 & -4 & 0 & 2 \\ 4 & 2 & 8 & 3 \end{bmatrix} \begin{array}{c} -6 \\ -4 \\ 2 \end{array}$ This game is strictly
$\qquad\; 5 \quad 2 \quad 9 \quad 3$ determined.   The (3, 2)
location is the saddle point.
The value of the game is 2.

3. (a) This game is not strictly determined, so there is no solution.

    (b) This game is strictly determined with value 4. The saddle point is at location (2,1). So the solution consists of the offense adopting strategy 2 and the defense adopting strategy 1.

5. (a) $E = [.3 \quad .7] \begin{bmatrix} 5 & 9 \\ 11 & 2 \end{bmatrix} \begin{bmatrix} .6 \\ .4 \end{bmatrix} = [9.2 \quad 4.1] \begin{bmatrix} .6 \\ .4 \end{bmatrix} = 7.16$

    (b) $E = [.5 \quad .5] \begin{bmatrix} -2 & 6 \\ 3 & 9 \end{bmatrix} \begin{bmatrix} .1 \\ .9 \end{bmatrix} = [.5 \quad 7.5] \begin{bmatrix} .1 \\ .9 \end{bmatrix} = 6.8$

    (c) $E = [.1 \;.4\; .5] \begin{bmatrix} -3 & 2 & 1 \\ 4 & -2 & 5 \\ 3 & 1 & 2 \end{bmatrix} \begin{bmatrix} .2 \\ .2 \\ .6 \end{bmatrix} = [2.8 \;-0.1\; 3.1] \begin{bmatrix} .2 \\ .2 \\ .6 \end{bmatrix} = 2.4$

7. $p_1 = \dfrac{210 - 175}{250 + 210 - 140 - 175} = \dfrac{35}{145} = 0.24 \qquad p_2 = 0.76$

    The farmer should plant 24% of the crop in the field and 76% in the greenhouse.

Chapter 10
Logic

1.   (a)   Statement. It is a true declarative sentence.
        (b)   Statement. It is a false declarative sentence.
        (c)   Not a statement. It is a question.
        (d)   Not a statement. It is an opinion.

3.   (a)   Statement. It is a true declarative sentence.
        (b)   Statement. It is a false declarative sentence.
        (c)   Statement. It is a true declarative sentence.
        (d)   Not a statement. It is a command.
        (e)   Not a statement. It is an opinion.

5.   (a)   Neither   (b) Disjunction
        (c)   Conjunction   (d) Disjunction

7.   (a)   Betty has blonde hair or Angela has dark hair.
        (b)   Angela does not have dark hair.
        (c)   Betty has blonde hair and Angela has dark hair.
        (d)   Betty does not have blonde hair and Angela does not have dark hair.

9.   (a) $p \wedge q$ (b) $p \vee q$ (c) $(\sim p) \wedge q$

11.   (a)   False because $4^2 = 15$ is false.
        (b)   False because the first part is false.
        (c)   True because both statements making up the conjunction are true.

13.   (a)   False because both parts are false.
        (b)   True because both parts are true.
        (c)   True because the first part is true.

15.   (a)   I do not have six one-dollar bills in my wallet.
        (b)   Roy cannot name all 50 states.
        (c)   A quorum was present for the meeting.

17.   (a)   I drink coffee at breakfast and I eat salad for lunch and I like a dessert after dinner.
        (b)   I drink coffee at breakfast, or I eat salad for lunch and I like a dessert after dinner.
        (c)   I drink coffee at breakfast and I eat salad for lunch, and I do not like a dessert after dinner.
        (d)   I drink coffee at breakfast and I eat salad for lunch, or I drink coffee at breakfast and I like dessert after dinner.

19.   (a)   T because all parts are T   (b)   F because $\sim p$ is F
        (c)   F because $\sim p$ is F   (d)   F because $\sim r$ is F
        (e)   T because $p \vee q$ is T

21.

p	q	$\sim p$	$\sim q$	$\sim p \wedge \sim q$
T	T	F	F	F
T	F	F	T	F
F	T	T	F	F
F	F	T	T	T

23. (a)   p: Jane brings chips.
              q: Tony brings drinks.
              r: Hob brings cookies.
              s: Ingred brings chips.
              t: Alex brings drinks.
              u: Hester brings drinks.
              v: Alice brings cookies.
              w: Jenn brings cookies.
              The statement is
              $(p \wedge q \wedge r) \vee (s \wedge (t \vee u) \wedge (v \vee w))$

    (b)   The statements have the truth values
              p = T, q = F, r = F, s = T, t = T, u = T, v = F, w = F
              Substituting these values in the symbolic statement we have
              $(T \wedge F \wedge F) \vee (T \wedge (T \vee T) \wedge (F \vee F))$
              which reduces to $F \vee (T \wedge T \wedge F)$ which is F.

Section 10.2

1. (a) If I have $5.00, then I can rent a video.
   (b) If I can rent a video, then I have $5.00.

3. (a) True because the hypothesis and the conclusion are both true.
   (b) True because the hypothesis is false.
   (c) False because the hypothesis is true and the conclusion is false.

5. Converse: If I live in Colorado, then I live in Denver.
   Inverse: If I do not live in Denver, then I do not live in Colorado.
   Contrapositive: If I do not live in Colorado, then I do not live in Denver.

7. (a) False because the components have different truth values, true and false, respectively.
   (b) True because both components are true.
   (c) True because both components are true.

9.

p	q	$p \wedge q$	$\sim(p \wedge q)$
T	T	T	F
T	F	F	T
F	T	F	T
F	F	F	T

11.

p	q	$\sim q$	$p \wedge \sim q$
T	T	F	F
T	F	T	T
F	T	F	F
F	F	T	F

13.

p	q	$\sim p$	$\sim q$	$\sim p \vee \sim q$
T	T	F	F	F
T	F	F	T	T
F	T	T	F	T
F	F	T	T	T

15.

p	q	q ∨ p	~p → (q ∨ p)
T	T	T	T
T	F	T	T
F	T	T	T
F	F	F	F

17.

p	q	r	p ∨ q	(p ∨ q) ∧ r
T	T	T	T	T
T	F	T	T	T
F	T	T	T	T
F	F	T	F	F
T	T	F	T	F
T	F	F	T	F
F	T	F	T	F
F	F	F	F	F

19.

p	q	r	p → q	q → r	(p → q) ∧ (q → r)
T	T	T	T	T	T
T	F	T	F	T	F
F	T	T	T	T	T
F	F	T	T	T	T
T	T	F	T	F	F
T	F	F	F	T	F
F	T	F	T	F	F
F	F	F	T	T	T

21.

p	q	r	p ∨ q	(p ∨ q) ↔ r
T	T	T	T	T
T	F	T	T	T
F	T	T	T	T
F	F	T	F	F
T	T	F	T	F
T	F	F	T	F
F	T	F	T	F
F	F	F	F	T
F	F	T	T	T

23.

p	~p	~(~p)	~(~p) ↔ p
T	F	T	T
F	T	F	T

25. (a)   p: your insurance company paid the provider directly for part of your expenses.
q: you paid only the amount that remained.
r: include on line 1 only the amount you paid.
The statement is (p ∧ q) → r.

     (b)   p: you leave line 65 blank.
q: the IRS will figure the penalty.
r: the IRS will send you the bill.
The statement is p → (q ∧ r).

     (c)   p: you changed your name because of marriage, divorce, etc.
q: you made estimated tax payments using your former name.
r: attach a statement to the front of Form 1040 explaining all the payments you and your spouse made in 1993.
s: attach the service center where you made the payments.
t: attach the name and SSN under which you made payments.
The statement is (p ∧ q) → (r ∧ s ∧ t).

1.

p	q	p → q	~q → ~p
T	T	T	T
T	F	F	F
F	T	T	T
F	F	T	T

Equivalent

3.

p	q	~p	p ∧ q	~p ∨ (p ∧ q)	~p ∨ q
T	T	F	T	T	T
T	F	F	F	F	F
F	T	T	F	T	T
F	F	T	F	T	T

Equivalent

5.

p	q	r	~p	q ∧ r	~p ∨ (q ∧ r)	p → (q ∧ r)
T	T	T	F	T	T	T
T	F	T	F	F	F	F
F	T	T	T	T	T	T
F	F	T	T	F	T	T
T	T	F	F	F	F	F
T	F	F	F	F	F	F
F	T	F	T	F	T	T
F	F	F	F	F	T	T

Equivalent

7.

p	q	p → q	~(p → q)	p
T	T	T	F	T
T	F	F	T	T
F	T	T	F	F
F	F	T	F	F

Not equivalent

9.

p	q	r	q ∨ r	p ∧ q	p ∧ (q ∨ r)	(p ∧ q) ∨ r
T	T	T	T	T	T	T
T	F	T	T	F	T	T
F	T	T	T	F	F	T
F	F	T	T	F	F	T
T	T	F	T	T	T	T
T	F	F	F	F	F	F
F	T	F	T	F	F	F
F	F	F	F	F	F	F

Not equivalent

11. Let p represent the statement "The exceptions above apply," and let q represent the statement "Use Form 2210." Then statement (a) can be represented by ~p → q and statement (b) can be represented by p ∨ q. Make a truth table and compare the truth values of ~p → q and p ∨ q.

p	q	~p	~p → q	p ∨ q
T	T	F	T	T
T	F	F	T	T
F	T	T	T	T
F	F	T	F	F

Since the truth values of ~p → q and p ∨ q are identical, the statements are equivalent.

13.   p: I will buy a jacket. q: I will buy a shirt. r: I will buy a tie.   Then statement (a) is p ∧ (q ∨ r)
(b) is (p ∧ q) ∨ (p ∧ r)
   We form a truth table and compare the truth values of the two statements.

p	q	r	q ∨ r	p ∧ (q ∨ r)	p ∧ q	p ∧ r	(p ∧ q) ∨ (p ∧ r)
T	T	T	T	T	T	T	T
T	F	T	T	T	F	T	T
F	T	T	T	F	F	F	F
F	F	T	T	F	F	F	F
T	T	F	T	T	T	F	T
T	F	F	F	F	F	F	F
F	T	F	T	F	F	F	F
F	F	F	F	F	F	F	F

Since the truth values of p ∧ (q ∨ r) and
(p ∧ q) ∨ (p ∧ r) are identical, they are equivalent.

Section   10.4

1.   p: Eat your beans.
      q: You may have dessert.
            p → q
            p
            ─────────
            q   Valid, Law of Detachment

3.   p: You do not study.
      q: You cannot do the homework.
      r: You cannot pass the course.
            p → q
            q → r
            ─────────
            p → r   Valid, syllogism

5.   p: You eat your beans.
      q: You may have dessert.
            p → q
            ~p
            ─────────
            ~q   Not valid.   See Example 5.

7.   p: The ice is six inches thick.
      q: Shelley will go skating.
            p → q
            ~q
            ─────────
            ~p
            Valid, Indirect reasoning

9.   p: Inflation increases.
      q: The price of new cars will increase.
      r: More people will buy used cars.
            p → q
            q → r
            ─────────
            p → r   Valid, Syllogism

11. Check [(p → q) ∧ (q ∧ r)] → (p ∨ r)

p	q	r	p→q	q∧r	p∨r	[(p→q) ∧ (q∧r)] → (p∨r)
T	T	T	T	T	T	T
T	F	T	F	F	T	T
F	T	T	T	T	T	T
F	F	T	T	F	T	T
T	T	F	T	F	T	T
T	F	F	F	F	T	T
F	T	F	T	F	F	T
F	F	F	T	F	F	T

Valid

13. Check [(p ∧ q) ∧ (p → ~q)] → (p ∧ ~q)

p	q	p∧q	p → ~q	p∧~q	[(p ∧ q) ∧ (p → ~q)] → (p ∧ ~q)
T	T	T	F	F	T
T	F	F	T	T	T
F	T	F	T	F	T
F	F	F	T	F	T

Valid

15. Check [(q → r) ∧ (~p ∨ q) ∧ p] → r

p	q	r	q→r	~p∨q	[(q→r) ∧ (~p∨q) ∧ p] → r
T	T	T	T	T	T
T	F	T	T	F	T
F	T	T	T	T	T
F	F	T	T	T	T
T	T	F	F	T	T
T	F	F	T	F	T
F	T	F	F	T	T
F	F	F	T	T	T

Valid

17. Check [(p → q) ∧ (p → r)] → (q ∧ r)

p	q	r	p→q	p→r	q∧r	[(p→q) ∧ (q→r)] → (q∧r)
T	T	T	T	T	T	T
T	F	T	F	T	F	T
F	T	T	T	T	T	T
F	F	T	T	T	F	F
T	T	F	T	F	F	T
T	F	F	F	F	F	T
F	T	F	T	T	F	F
F	F	F	T	T	F	F

Not valid

19. Check [(p → q) ∧ (q → r) ∧ ~q]→ ~r

p	q	r	p→q	q→r	~q	~r	[(p→q) ∧ (q→r) ∧ ~q]→ ~r
T	T	T	T	T	F	F	T
T	F	T	F	T	T	F	T
F	T	T	T	T	F	F	T
F	F	T	T	T	T	F	F
T	T	F	T	F	F	T	T
T	F	F	F	T	T	T	T
F	T	F	T	F	F	T	T
F	F	F	T	T	T	T	T

Not valid

21.   p: I mow the lawn.
q: I trim the hedge.
r: I may go to the movie.
The argument can be represented as
Premise: $(p \wedge q) \rightarrow r$
                          q_____
Conclusion: $p \rightarrow r$
Let's form the truth table and see if
$\{[(p \wedge q) \rightarrow r] \wedge q\} \rightarrow (p \rightarrow r)$ is a tautology (true for all values
of p, q, and r).

p	q	r	p∧q	(p∧q)→r	p→r	{[(p∧q)→r]∧q}→(p→r)
T	T	T	T	T	T	T
T	F	T	F	T	T	T
F	T	T	F	T	T	T
F	F	T	F	T	T	T
T	T	F	T	F	F	T
T	F	F	F	T	F	T
F	T	F	F	T	T	T
F	F	F	F	T	T	T

Since the last column has all T, the argument is valid.

23.   This is of the form
              $p \vee q$
              ~ p_____
                q
This form is the disjunctive syllogism and so it is valid.

25.   This argument is of the form
              $p \rightarrow q$
              ~ q_____
              ~ p
This is indirect reasoning, so it is valid.

27.   This argument is of the form
              $p \vee q$
              ~ p_____
                q
This is a disjunctive syllogism, so it is valid.

Review Exercises,   Chapter 10

1.   (a)   Statement   (b)   Not a statement
      (c)   Not a Statement   (d)   Statement

3.   (a)   Rhonda is not sick today.
      (b)   Rhonda is sick today and she has a temperature.
      (c)   Rhonda is not sick today and Rhonda does not have a
            temperature.
      (d)   Rhonda is sick today or she has a temperature.

5.   (a)   True   (b)   True   (c)   False   (d)   True

7.   (a)   T   (b)   T   (c)   T   (d)   F

9.   (a)   True   (b)   False

11. *Inverse*: "If I do not turn my paper in late, then I will not be penalized."
    Converse: "If I will be penalized, then I will turn my paper in late."
    Contrapositive: "If I will not be penalized, then I will not turn my paper in late."

13.

p	q	~p	p∧q	~p → (p∧q)
T	T	F	T	T
T	F	F	F	T
F	T	T	F	F
F	F	T	F	F

15.

p	q	p∨q	p∧(p∨q)
T	T	T	T
T	F	T	T
F	T	T	F
F	F	F	F

Since p and p∧(p ∨ q) have the same truth values they are logically equivalent.

17. This argument is of the form

$$p \rightarrow q$$
$$\underline{\sim q}$$
$$\sim p$$

This is indirect reasoning, so it is valid.

19. This is an argument of the form

$$p \rightarrow q$$
$$\underline{q \rightarrow r}$$
$$p \rightarrow r$$

so it is valid by syllogism.

21. This argument is of the form

$$p \vee q$$
$$\underline{\sim p}$$
$$q$$

This form is a disjunctive syllogism so it is valid.

Appendix A
Algebra Review

Section A.1

1. $(-1)13 = -13$ 3. $-(-23) = 23$

5. $(-5)(6) = -30$ 7. $5(-7) = -35$

9. $-(7 - 2) = -5$ 11. $21/(-3) = -7$

13. $(-4) + (-6) = -10$ 15. $(-4)(2) = -8$

17. $5/3 + 4/3 = 9/3 = 3$ 19. $12/5 - 3/5 = 9/5$

21. $2/3 + 3/4 = 8/12 + 9/12 = 17/12$

23. $5/6 - 7/4 = 10/12 - 21/12 = -11/12$

25. $2/5 + 1/4 = 8/20 + 5/20 = 13/20$

27. $4/7 - 3/5 = 20/35 - 21/35 = -1/35$

29. $(3/4)/(9/8) = (3/4) \times (8/9) = 2/3$

31. $(2/7)/(4/5) = (2/7) \times (5/4) = 5/14$

33. $(1/3)(1/5) = 1/15$ 35. $2/5 \times 4/3 = 8/15$

37. $(-3/5)(-4/7) = 12/35$

39. $3/11 + 1/3 = 9/33 + 11/33 = 20/33$

41. $5/7 \div 15/28 = (5/7) \times (28/15) = 4/3$

43. $(5/8) \div (1/3) = (5/8)(3) = 15/8$

45. $(3/4 + 1/5) \div (2/9) = (15/20 + 4/20)(9/2) = (19/20)(9/2) = 171/40$

47. $-2(3a + 11b) = -6a - 22b$ 49. $-5(2a + 10b) = -10a - 50b$

Section A.2

1. $2x - 4 = -10$
    $x = -3$ is a solution, since $2(-3) - 4 = -6 - 4 = -10$

3. $2x - 3 = 5$ 5. $4x - 3 = 5$
    $2x = 8$ $4x = 8$
    $x = 4$ $x = 2$

7. $7x + 2 = 3x + 4$ 9. $12x + 21 = 0$
    $4x = 2$ $12x = -21$
    $x = 1/2$ $x = -21/12 = -7/4$

11. $3(x - 5) + 4(2x + 1) = 9$
     $3x - 15 + 8x + 4 = 9$
     $11x = 20$
     $x = 20/11$

13. $\dfrac{2x + 3}{3} + \dfrac{5x - 1}{4} = 2$
     $8x + 12 + 15x - 3 = 24$
     $23x = 15$
     $x = \dfrac{15}{23}$

15. $\dfrac{12x + 4}{2x + 7} = 4$
     $12x + 4 = 8x + 28$
     $4x = 24$
     $x = 6$

17. (a) $y = 0.20(650) + 112 = \$242$
     (b) $y = 0.20(1500) + 112 = \$412$
     (c) Solve $0.20x + 112 = 302$
          $0.20x = 190$
          $x = 950$ miles

19. (a) $y = 0.42(42 - 8) = 0.42(34) = \$14.28$
     (b) $y = 0.42(113 - 8) = 0.42(105) = \$44.10$
     (c) Solve $0.42(x - 8) = 22.26$
          $x - 8 = 53$
          $x = 61$ pounds

Section A.3

1.

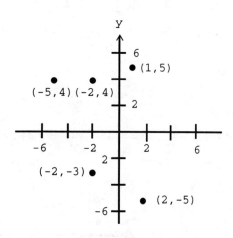

3.

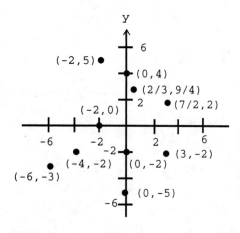

5. (a) Second quadrant: negative x-coordinates,
 positive y-coordinates
 (b) Third quadrant: negative x-coordinates,
 negative y-coordinates
 (c) Fourth quadrant: positive x-coordinates,
 negative y-coordinates

7.

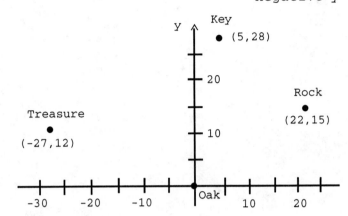

Section A.4

1. (a) 9 > 3 is True, since 9 - 3 = 6.
 (b) 4 > 0 is True, since 4 - 0 = 4.
 (c) -5 > 0 is False, since -5 - 0 = -5.
 (d) -3 > -15 is True, since -3 - (-15) = 12.
 (e) 5/6 > 2/3 is True, since 5/6 - 2/3 = 5/6 - 4/6 = 1/6.

3. 3x - 5 < x + 4
 2x < 9
 x < 9/2

5. 5x - 22 ≤ 7x + 10
 -2x ≤ 32
 x ≥ -16

7. 3(2x + 1) < 9x + 12
 6x + 3 < 9x + 12
 -3x < 9
 x > -3

9. 3x + 2 ≤ 4x - 3
 -x ≤ -5
 x ≥ 5

 ●————————→
  5

11. 6x + 5 < 5x - 4
 x < -9

 ←————○——————
   -9

13. 3(x + 4) < 2(x - 3) + 14
 3x + 12 < 2x - 6 + 14
 x < -4

 ←————○——————
   -4

15. 3(2x + 1) < -1(3x - 10)
 6x + 3 < -3x + 10
 9x < 7
 x < 7/9

 ←————○——————
   7/9

17. -16 < 3x + 5 < 22
 -21 < 3x < 17
 -7 < x < 17/3

 ——○————————○——
  -7   17/3

195

19. $14 < 3x + 8 < 32$
        $6 < 3x < 24$
        $2 < x < 8$

21. $3x + 4 \leq 1$
        $3x \leq -3$
        $x \leq -1$

        $(-\infty, -1]$

23. $-7x + 4 \geq 2x + 3$
        $-9x \geq -1$
        $x \leq 1/9$

        $(-\infty, 1/9]$

25. $-45 < 4x + 7 \leq -10$
        $-52 < 4x \leq -17$
        $-13 < x \leq -17/4$
        $(-13, -17/4]$

27. $\dfrac{6x + 5}{-2} \geq \dfrac{4x - 3}{5}$
        $30x + 25 \leq -8x + 6$
        $38x \leq -19$
        $x \leq -1/2$

29. $\dfrac{2}{3} < \dfrac{x + 5}{-4} \leq \dfrac{3}{2}$
        $-8 > 3x + 15 \geq -18$
        $-23 > 3x \geq -33$
        $-23/3 > x \geq -11$

31. $75 \leq 35 + 5x < 90$
        $40 \leq 5x < 55$
        $8 \leq x < 11$
        8, 9, or 10 correct answers

33. $85 \leq 3x + 25 \leq 100$
        $60 \leq 3x \leq 75$
        $20 \leq x \leq 25$
        20, 21, 22, 23, 24, or 25